To Steve
Happy Birthday
, 1990

Jeff

WARBIRDS

WARBIRDS

CLASSIC AMERICAN FIGHTERS & BOMBERS

Michael O'Leary,
Norman Pealing & Mike Jerram

MILITARY PRESS
New York

Compiled, edited and designed by Richard and Janette Widdows

[The material in this book previously appeared in the Osprey Aerospace publications *Gunfighters* and *Bombing Iron* (Michael O'Leary), *American Warbirds* (Norman Pealing) and *Warbirds* (Mike Jerram)]

CONTENTS

MICHAEL O'LEARY has been employed for a number of years as editor and associate publisher for a large group of aviation magazines. Based in Los Angeles, he has had the opportunity to photograph many different types of aircraft–from the newest high-tech warplanes to primitive contraptions from the dawn of flight. However, his favorite aircraft remain those classic machines from World War 2–the warbirds, the ultimate piston-engined planes. Having photographed several hundred of these in flight, he's always attracted by the grace and symmetry of what were once deadly weapons of war.

Some of his pictures in this book were taken from the drafty, vibrating interior of the great old Mitchell *Executive Sweet*, photographers slowly being deafened from the immense roar of the might Wright R-2600 radials. However, a number of other camera-ships were used by Michael: the Beechcraft F33A Bonanza, Beechcraft T-34 Mentor, Beechcraft C-45 Expeditor, North American P-51B and P-51D Mustang, Piper Lance and Piper Seneca. He is grateful for the skill of their pilots and thanks go especially to Bruce Guberman, who flew the majority of his photo-missions in this book.

NORMAN PEALING began taking pictures before he entered the Royal Air Force in 1958, but his photographic portfolio was not allowed to expand into aviation subjects in the days when RAF Marham and Wittering were stuffed full of Valiant nuclear bombers.

In 1965 he joined the British Aircraft Corporation (BAC), and began making sales and publicity films to promote all the company's products, which included guided weapons, satellites and military and civil aircraft. He attended many first flights and took part in the demonstration tours of the One-Eleven and Concorde airliners.

By 1983 he chose to leave what had become the Weybridge Division of British Aerospace (BAe) to form his own company at Fairoaks Airport in Surrey. Aviation Image specializes in aviation photography and film and video production for advertising, sales support, public relations, publishing and television requirements. Like his two co-photographers on this book, Norman has taken the pictures and written the captions for a number of successful titles on aviation in the Osprey Color Series.

MIKE JERRAM's earliest recollection of warbirds is of watching Fleet Air Arm Fireflies, Sea Furies and Sea Hornets landing at Lee-on-Solent in Hampshire, England. Although his work as a freelance aviation writer and photographer now brings him into contact with every kind of flying machine from microlights to the Space Shuttle, he has a particular passion for vintage aircraft and warbirds and has spent many years photographing them, not entirely without incident. Two trips in a B-17G Flying Fortress were both punctuated by tailwheel failures, the first when the bomber was at the end of its display slot and the RAF Red Arrows display team were due on next. Mike was delegated to go back to the tail cone to investigate the recalcitrant wheel's apparent failure to extend. No-one told him they carried a spare wheel back there . . .

On another occasion, while shooting close to the active runway at an American airshow, he committed the cardinal sin of the action photographer and turned his back on his subjects to reload cameras. A sudden change in engine note and a screech of protesting rubber focused attention, but not cameras, just as the Corsair pilot managed to tame his ground loop a few yards from the petrified photographer.

Mike lives and works on the south coast of England. He is contributing editor of the British general aviation magazine *Pilot* and an assistant compiler of *Jane's All the World's Aircraft*.

INTRODUCTION

The vast majority of the historic warplanes from World War 2 were junked by the thousand in peacetime, some types lost forever to the breakers' cutting torches. Happily there is today a greater awareness of the value of such treasures, and that awareness is being channeled not just to static preservation in museums, laudable though many of them are, but to the maintenance and creation of airworthy aircraft – warbirds.

What is a warbird? There is no positive definition, but broadly speaking the term is applied to aircraft which served in World War 2 or in the immediate postwar period, including the Korean War. In this book you will find some aircraft which stretch that definition a little, others which played supporting roles in battle but had neither the weaponry nor stamina to act as aggressors. But they are all

part of the warbird movement, and welcome. Few but the best-heeled can aspire to own and operate a Mustang or a Bearcat, whose price tags (when they do come on the market) run to six figures. But a Piper Cub costs no more, and often less, than a saloon car, and has equal claim to wear a uniform.

It's really by chance that the few World War 2 bombers still flying today survived the mass aircraft scrappings following the conclusion of the war, a time when America was trying to convert its swords into ploughshares. Perhaps it's only in America that these aircraft could have survived. Some companies had a need for four-engined bombers (photo-mapping platforms, cargo-haulers or gun-runners), while the medium bombers found new careers in aerial applications, fire bombing and, later, conversion into high-

speed executive and business transports.

The task of restoring a vintage aircraft and bringing it back to wartime configuration is daunting at best. The availability of original parts such as turrets has become difficult (not to mention expensive), while other problems such as liability insurance and the decreasing availability of avgas are major issues facing the warbird community.

Yet thanks to dedicated groups such as Warbirds of America and the Confederate Air Force, the warbird movement is thriving. Long-forgotten aircraft are being sought out in the obscurer parts of the world, where they were cast off by air forces on the way to better things, and painstakingly restored to their rightful place—in the air. If you wonder why, take a careful look at the crowd at the next major airshow you visit. While some people will be captivated by the deafening afterburner roar and now-you-see-it-now-you-don't lightning dashes of modern jet fighters, and others will dwell on every twist and turn of

buzzing, bee-like Pitts Specials, the only sight and sound *guaranteed* to get everyone's head out of their program or feet out of the refreshment tent is that of a World War 2 fighter or bomber. It's more than nostalgia, for many of those so moved have parents too young to remember the days when these aircraft were the last word in military sophistication.

And that, perhaps, is the key to the warbird's charisma. It represents the combat aircraft and pilot at their purest, man and machine unfettered by electronic gadgetry of modern push-button fire-and-forget warfare. In these warplanes you really *can* get to see the whites of the pilot's eyes.

Currently, many fine examples of restored World War 2 and Korean War aircraft grace our skies. This book attempts to illustrate most of them and recall an era in history whose lessons, hopefully, will not be forgotten. The contributors trust that you, the reader, will enjoy it.

Wild horses:
MUSTANG

One of the first fighters in history and arguably the most outstanding combat aircraft to emerge from World War 2, the North American P-51 Mustang was the breathtaking result of an April 1940 agreement between the British Air Purchasing Commission and the great 'Dutch' Kindelberger, chairman of North American Aviation, for the design and development of a new fighter for the RAF. NAA took just 117 days to design and build the NA-73X prototype, which made its maiden flight on 26 October 1940 following a six-week delay while Allison struggled to deliver their new V-1710 vee-12 engine. The rest, as they say, is history.

LEFT Stephen Grey puts his P-51D *Moose* through its paces in the vicinity of North Weald in the South of England during the 1986 'Fighter Meet'. The same aircraft is pictured in profile during 'escort duty' on page 220. [*Norman Pealing*]

ABOVE Phoenix from the ashes. Well, not quite ashes – North American P-51D-25-NA, United States Army Air Force (USAAF) serial number 44-73264, was nearly written off in September 1981 when it was caught by a gust of wind landing at Omaha, Nebraska.

The airplane cartwheeled, resulting in extremely heavy damage. Utilizing many new and rebuilt parts (including a new fuselage), N5428V was eventually brought back to life over a period of several years. [*Michael O'Leary*]

RIGHT With the Rolls-Royce/Packard V-1650 Merlin ticking over, the pilot advances the throttle just a hair to get the Mustang moving towards the taxiway. Bearing the famous insignia of the 4th Fighter Group, P-51D-25-NT N2869D, s/n 44-84390, survived a long and rigorous career as an unlimited air racer. The Packard 1590 hp V-1650-7 Merlin-engined D model was by far the most numerous version of the Mustang, outnumbering all the other models combined. A grand total of 15,586 Mustangs were delivered. [*Michael O'Leary*]

ABOVE Nose art was by no means confined to bombers, nor only used as a reminder of the crew's more basic needs and fantasies! *Death Rattler* represented a truism in Mustang terms – many a German and Japanese pilot came to rue the day he mixed it with a P-51. [*Norman Pealing*]

BELOW Three Merlins and an Allison: P-51Ds follow a P-40 Warhawk as they thread their way along the taxiway at Harlingen, the sound of their vee-12 engines sweet music to thousands of spectators during the 1987 Confederate Air Force display. P-40F/Ls were also powered by Packard Merlins. General Eisenhower was so impressed with the Mustang that he flew over the Normandy beachhead in a specially modified example flown by Major General Elwood 'Pete' Quesada. [*Norman Pealing*]

ABOVE Sporting distinctive Normandy invasion stripes, this view of a P-51D – the very one flown at Harlingen by a certain Brigadier-General 'Chuck' Yeager – shows off its big four-bladed, blunt-tipped Hamilton Standard propeller. The D model had a top speed of 437 mph and packed six .50 calibre Browning MG53-2 machine-guns in the wings, each fed by 270- or 400-round magazines. [*Norman Pealing*]

ABOVE Presumably a checkerboard tailplane on your P-51 comes in handy for a quick game of chess during those long hours at dispersal. The Mustang was not a comfortable aircraft to fly on those daunting escort missions and the cockpit was noisy and cold at extreme altitudes. Yet few would have changed it for anything other than a later mark—except the squirrely lightweight H model. [*Norman Pealing*]

P-51 MUSTANG

Beauty and the beast. Pete Regina's authentically restored combat P-51B N51PR in formation during the August 1985 Gathering of Warbirds airshow in Madera, California, an event that draws some of the finest warbirds on the West Coast. Pete's wingman is Commonwealth Aircraft Corporation CA-18

Mustang Mk 23 A68-198. Now registered N286JB, the Australian-built Mustang has had its aluminum skin highly polished and is painted in post-World War 2 markings. N286JB last saw active service with the Royal Australian Air Force and then served a long period of flogging around the pylons at Reno as an

unlimited air racer. The Mustang then went on to become a trainer for Piper's ill-fated Enforcer program before being brought back into basically stock condition. [*Michael O'Leary*]

LEFT One of the rarest of all Mustang variants is the A-36A dive-bomber. Originally named Invader, the A-36A was derived from the early P-51A airframe and optimized for the dive-bombing role, including the addition of dive-brakes in the wing. The A-36A, which later reverted to the Mustang name, was a spectacular combat machine. Extremely fast down low (the A-36A was powered by the Allison engine which, due to a lack of efficient supercharging, lost out at high altitude), and more than capable of defending itself against enemy fighters, it racked up many successes and remained active until the end of the war. Only 500 were built and most of these were expended in combat. Fortunately, A-36A-1-NA s/n 42-83731 survived to be restored by Dick Martin for owner Tom Friedkin. The plane is seen nearing the end of its restoration at Palomar, California. [*Michael O'Leary*]

BELOW AND RIGHT Few families can boast two brothers who are both Mustang rebuilders and owners. Pete and Angelo Regina both desired Mustangs but set out in different ways to obtain their own fighter. Pete discovered a rare P-51B wing, then obtained a fuselage from an Israeli Air Force P-51D and built all the missing parts. Brother Angelo, a Flying Tiger Airlines pilot, discovered a P-51D being used as a playground toy in an Israeli kibbutz. Both restorations have now been purchased by Joe Kasperoff. [*Michael O'Leary*]

Mustangs became the last operational propeller-driven fighters from World War 2 for three basic reasons: availability of airframes (many Mustangs had been stored); high performance; and ease of maintenance in the field compared to other American fighters like the Thunderbolt and Lightning.

ABOVE Many parts for Pete Regina's restoration had to be manufactured from scratch, giving some idea of the amount of money and dedication expended by the owners of these classic machines. Now registered N251A, the A-36A can be seen at West Coast airshows. [*Michael O'Leary*]

P-51 MUSTANG

High over a wind-tossed sea, Hess Bomberger formates P-51D-30-NA USAAF s/n 44-74497. A World War 2 Mustang combat pilot, Hess has painted his mount in the markings of one of the aircraft he flew. N6320T last saw active service with the Royal Canadian Air Force. [*Michael O'Leary*]

P-51 MUSTANG

BELOW Mustang C-FBAU, USAAF s/n 44-73140, was destroyed after a deadstick landing in 1984. [*Michael O'Leary*]

BOTTOM Painted by a setting Florida sun, Bob Pond flies his P-51D N151BP. It's finished in 361st Fighter Group colors and accompanied here by his

Grumman/Eastern TBM-3E Avenger in World War 2 Fleet Air Arm markings. Looking at first like a Cavalier conversion (note the 12-inch fin cap atop the vertical tail) this aircraft was actually a 'homebuilt' Cavalier – being so modified by a private owner during the 1960s and even incorporating such 'updates' as air conditioning! [*Michael O'Leary*]

BELOW This Canadian Mustang is just settling on its landing gear, the Merlin giving its characteristic snap-crackle-pop as it is throttled back. The Royal Canadian Air Force equipped with P-51Ds in 1947, and few them with auxiliary squadrons until 1956. [*Mike Jerram*]

BOTTOM Canadian duo. Surplus Mustangs from the Royal Canadian Air Force bolstered the American supply of P-51Ds during the late 1950s and early 1960s. Canada was soon stripped of the fighters, but recently more civil Mustangs have been brought back into Canada to feed the growing collectors' appetite for the machine. Ross Grady is lead with his Cavalier-modified Mustang C-GMUS. This aircraft was obtained surplus from the air force of Bolivia, where it had operated as FAB-523. Bolivia was supplied with a small number of Cavalier F-51D Mk 2 Mustangs during the late 1960s as part of the US Government's Project Peace Eagle (an incongruous name, considering the mission). C-GMUS was even assigned a new USAF serial, 67-22581, thus losing its original wartime identity. Ross's wingman, Richie Rasmussen's C-GRLR (rebuilt from the badly damaged N5471V), was recently imported back into the Unites States for a new owner. [*Michael O'Leary*]

BELOW Lockheed test pilot Skip Holm demonstrates the Mustang's gear sequencing on Jim Beasley's NL51JB. Note how the clam shell doors open to allow the gear to drop, the doors then closing to permit minimum interference of cooling air to the radiator. [*Michael O'Leary*]

RIGHT Against a troubled sea, Holm makes a high-speed pass in *Bald Eagle*. A pilot's Mustang formation skills really show when flying with a much slower camera ship—in this instance a Beechcraft Bonanza. [*Michael O'Leary*]

P-51 MUSTANG

BELOW *Old Crow* crackles past, showing to good effect the Mustang's wide track landing gear made possible by the neat inward-retracting gears. This is obviously a much better arrangement than the outward-retracting gears found on aircraft like the Spitfire and Bf 109, whose pilots had to contend with a very narrow main track which hardly helped to contain the swing caused by powerful engines on take-off and landing; on some of the more rudimentary front-line airfields taxiing was distinctly taxing. [*Norman Pealing*]

RIGHT Bob Byrne and his gaily-painted *Rascal* are regular visitors to many American airshows. *Rascal* is another ex-Royal Canadian Air Force Mustang (RCAF 9270) and saw service with the USAAF as s/n 44-74774. *Rascal* was Byrne's first Mustang, but the '51 bug has bit hard and he has purchased several other Mustangs – including a very rare dual control TF-51D. [*Michael O'Leary*]

BELOW RIGHT The powerful lines of the Mustang are well portrayed in this taxi shot of Don Davidson's beautiful *Double Trouble Two*. Davidson equipped N51EA (s/n 44-63507) with a modern IFR panel, including autopilot and LORAN-C for pin-point navigation. His 100-plus hours of instrument time in the P-51 is probably a record! [*Michael O'Leary*]

ABOVE, LEFT AND RIGHT Canadian Warbird personality Jerry Janes has long been interested in high performance aircraft. Accordingly, during the 1970s, he had a P-51D restored from the ground up; the result is not only well-suited for airshows but also a fairly practical cross-country machine. Painted in desert camouflage, C-GJCJ is being flown (left) by Merlin engine rebuilder Dave Zeuschel over Southern California. C-GJCJ is now owned by David Price and registered N51D. The distinctive radiator chin beneath the fuselage is clearly visible in this view. The small flap at the rear of the bulge is open, allowing air to rush over the cooling gills during the vital moments after take-off. [*Michael O'Leary*]

BELOW The customized interior of David Price's N51D is typical of many refurbished Mustangs now flying, the original pilot's tub being plenty big enough to fit another seat in behind the driver. [*Michael O'Leary*]

RIGHT *Cottonmouth* is finished in the colors of a Mustang IV (the British Commonwealth always insisted on their own designations for American-built aircraft) serving with No 3 Squadron Royal Australian Air Force operating out of Zarro, Yugoslavia at the end of World War 2. The light blue fin decorated with the five stars of the Southern Cross is correct, but the shade of brown and mustard used is a bit off, to put it politely. [*Mike Jerram*]

BELOW Mustang at the gallop: *Passion Wagon* on the roll with the Merlin producing 1490 hp at 61 inches of boost and the 11-foot Hamilton Standard paddle-bladed propeller tearing out chunks of air at 3000 rpm. [*Mike Jerram*]

ABOVE Heading a stunning line-up of beautifully restored P-51Ds, and equipped with suitably decorated nose, *Passion Wagon* represents an aircraft of the 357th Fighter Group, who were based in Britain during 1944–45. The checkered nose cowl and striped spinner were synonymous with this group, as was the overall Olive Drab/Medium Green finish of the unit's Mustangs. Eventually the group dispensed with the camouflage green and the majority of 357th FG aircraft spent most of the late war period in dazzling polished metal. [*Mike Jerram*]

LEFT Mission completed. *Passion Wagon* rests as the sun goes down on another day in the long life of a fighter whose inauspicious beginnings could scarcely have heralded the Mustang's eventual emergence as the best Allied combat aircraft of World War 2. [*Mike Jerram*]

P-51 MUSTANG

Mustang air power at its very best: Jeff Ethell flies Mike Clark's P-51D N1451D *Unruly Julie* while Robb Satterfield formates in his N7722C (s/n 44-73420) *Miss Torque*. The checkered nose of the closest P-51 is reminiscent of the colors worn by the Mustangs of the 78th Fighter Group, a unit based in England during the last 18 months of the war. The plumbed wing hardpoints of the P-51 were a boon to fighter and bomber squadrons alike because it meant the aircraft's overall range could be virtually doubled.

Carrying a pair of standard 75 gallon drop tanks, *Untruly Julie*'s legs are considerably longer in this configuration than *Miss Torque*'s. [*Michael O'Leary*]

LEFT Stars and stripes aloft, Bill Clark sits proudly at the helm of his P-51D *Dolly*. [*Mike Jerram*]

RIGHT Off and running, John Baugh's P-51D *Miss Coronado* was once the mount of famed North American Rockwell test pilot and airshow performer Bob Hoover. Removal of fuselage fuel tank, armor plate and military radio gear leaves room aplenty for two under the Mustang's bubble top canopy. [*Mike Jerram*]

BELOW Mustang: the fiery wild horse of the American prairie. The P-51D pictured here was the definitive Mustang, powered by a Packard-built Rolls-Royce Merlin engine delivering 1695 hp. Earlier Allison engined-Mustangs lacked puff at high altitude and were relegated to tactical reconnaissance duties with the RAF, but the P-51D could top 41,000 feet and reach 438 mph. With drop tanks the Mustang's fuel load gave it very long legs, enabling it to fly bomber escort on round-trip missions of 1500 miles and more. [*Mike Jerram*]

ABOVE Bill Clark again, this time squeaking *Dolly* onto the runway at Oshkosh, Wisconsin for a one-wheel landing at the end of a Warbirds of America mission. [*Mike Jerram*]

BELOW AND RIGHT Like their four-legged counterparts P-51s need plenty of grooming, but at least they don't kick. Scott Smith's P-51D is as slick as they come, with wet-look black and white paint and . . . a red leather interior. He calls it *Ge Ge II*. Some horse! [*Mike Jerram*]

ABOVE Brightly painted *Stump Jumper* seen low over the Texas countryside during June 1983 while being piloted by owner Jerry Hayes. S/n 45-11367 has an interesting history. Last serving in the West Virginia Air National Guard (the last ANG unit to operate the Mustang), the fighter was pensioned off at Norton AFB in Southern California where it became part of a small museum. A new base commander, having little use for 'old' aircraft, ordered the museum machines scrapped. Fortunately Ed Maloney, owner and operator of The Planes of Fame Air Museum, rescued it. Years later, after another Mustang became fully operational, 11367 was sold to raise money for other projects. After stripping off layers of old paint, B-52 pilot Robin Collard uncovered the name *Stump Jumper* – apparently relating to a forced landing while serving with the West Virginia ANG.
[*Michael O'Leary*]

Los Angeles newspaper magnate Howard Keefe races his clip-wing P-51D *Miss America*. His Packard-Merlin has been tweaked up to produce more than 2000 hp. Note also the Hoerner wingtips and low profile tinted canopy. *Miss America* caused hearts to flutter in more ways than one in 1970 when a runaway rudder trim tab jammed hard over as Keefe was boring down the straightaway at Reno at 450 mph

during the Harrah's Trophy unlimited race. He chopped the power as the Mustang skidded towards the grandstand and managed to bring the runaway to heel. [*Mike Jerram*]

LEFT 'Pope Paul I', alias Experimental Aircraft Association president Paul Poberezny, flies this Cavalier conversion of the P-51D at airshows and on business trips for the EAA. Cavalier Corporation of Sarasota, Florida specialized in civilian conversions of P-51Ds, outfitting them with two-seat cockpits with modern instrument panels and avionics, baggage lockers in the wing gun/ammunition bays and a wide choice of fuel tankage offering a range of up to 2000 miles. At 400 mph-plus the Mustang is a swift, if less than roomy cross-country carriage for those who like to arrive with more style than your mass-production Wichita wonder can offer. Poberezny's tall-finned Cavalier Mustang wears the distinctive color scheme of *Lou IV*, a P-51D flown by 361st Fighter Group commander Colonel T.J.J. Christian Jr. The 361st

called themselves *The Yellowjackets*, hence the yellow cowling and spinner. Until recently the EAA also flew the oldest surviving Mustang, a prototype XP-51 Apache dating from 1941 which was the fourth aircraft built by North American. Now retired from active duty, it's displayed at the EAA Museum of Flight at Oshkosh. [*Mike Jerram*]

BELOW Latin America has been a rewarding hunting ground for warbird collectors and Mustang-lovers in particular, since many small air forces continued to operate them into the 1970s and one – the Dominican Air Force – still flies P-51Ds. *El Gato Rapido* came from the Costa Rican Air Force in 1978 and is pictured here after arrival in the United States. [*Mike Jerram*]

LEFT Six .50 machine-guns were standard P-51D armament and used to good effect by Lieutenant Claiborne H. Kinnard Jr of Franklin, Tennessee, who became an eight-victory ace while flying the original *Man O'War* with the 4th Fighter Group of the Eighth Air Force in England. The latter-day *Man O'War* has 16 swastika 'kill' markings along the canopy tail. [*Mike Jerram*]

LEFT North American developed a lightweight version of the Mustang in an attempt to improve high-speed performance and manoeuvrability without employing a bigger engine. The production version of the lightweight craft was designated P-51H. With a water-injected Packard V-1650-9 engine putting out a maximum 2218 hp, the '51H could reach 487 mph at 25,000 feet and was the fastest of all production variants. P-51Hs, identifiable by the deeper, slightly longer fuselage, taller fin and rudder and small wheels, saw only limited war service in the Pacific shortly before the Japanese surrender, but were flown by USAF and Air National Guard units in peacetime.

Some 555 were built, but few survive. This rare airworthy example is finished in the colors of the single aircraft supplied to the Royal Air Force for evaluation. The remarkable development of the P-51 is no better illustrated than by the P-51H's performance – fully 100 mph faster than the first Mustangs, which predated it by little more than three years. [*Mike Jerram*]

BELOW The P-51D introduced a beautiful teardrop hood which transformed visibility from the cockpit. Many earlier models were refitted with a bulged sliding Malcolm canopy. [*Norman Pealing*]

ABOVE The P-51D's clear bubble canopy was one of the first of its kind for fighter aircraft, developed to improve combat visibility. Earlier P-51B and 'C models had heavy-framed 'razorback' canopies which severely limited the pilot's vision, especially to the rear. The British Malcolm hood went part-way to solving the problem, but North American Aviation drew on moulding techniques used for manufacturing glazed nose transparencies for bombers to develop the bubble top capable of withstanding flight speeds of approaching 500 mph and rapid temperature changes during combat manouevring from high altitude. Pilots loved the bubble's near-perfect vision, but were less pleased that it turned the cockpit into a hothouse in bright sunlight. [*Mike Jerram*]

LEFT Another design masterpiece from North American Aviation, the F-82 (post-1948 designation) Twin Mustang would have escorted the high-flying B-29s all the way from their bases in China and the Pacific islands to targets in Japan and back again had the war continued beyond 1945. After the A-bomb attacks on Hiroshima and Nagasaki the Army Air Force promptly canceled some 480 of the 500 P-82Bs on order. Created by joining two P-51 fuselages at the wings and tailplane, the prototype XP-82 made its maiden flight on 15 April 1945. The AAF subsequently secured a trickle of postwar deliveries, production reaching the 400-mark with the advent of the F-82G night-fighter equipped with SCR-720 search radar. [*Norman Pealing*]

RIGHT The P-82E day escort model (100 built) also had a useful attack capability, armed with six .50 caliber machine-guns in a detachable tray, plus a 4000-lb bomb load or 24 5-inch air-to-surface rockets.

The Twin Mustang saw considerable action in Korea, where Lt Hudson of the 68th Fighter Squadron in his P-82B gained the distinction of scoring the first kill of that conflict (also the first Air Force victory). Most Twin Mustangs served with Air Defense Command. The P-82 seen here was in airworthy condition when it attended the 1987 Confederate Air Force airshow, but a landing accident after its display caused serious damage to the propellers, landing gears and airframe. Fortunately the crew were able to walk away from the crash and the owners have pledged to restore this aircraft to flying condition. [*Norman Pealing*]

BELOW A firm favorite with Confederate Air Force enthusiasts for over two decades, *Old Red Nose* has been part of the amazing warbird scene at Harlingen since the mid-1960s. Originally a USAAF Mustang built in 1944, this P-510-20-NA was then transfered to the Royal Canadian Air Force before appearing on the civil register in the 1950s. [*Mike Jerram*]

Bird of prey: WARHAWK

A prewar design, the Curtiss P-40 series fought on every combat front and flew right until the end of the war, compiling an impressive string of victories. Although generally outclassed by enemy fighters, the Warhawk family (the name Kittyhawk was only applied to aircraft delivered to the RAF), proved particularly adept in the close support role. Aggressively flown by RAF pilots in the Western Desert, the P-40 was a constant threat to German motorized columns and troops in the open. Nearly 14,000 Warhawks had been produced when the final delivery (a P-40N) took place in March 1944.

LEFT Hickham Field, Hawaii, 7 December 1941: USAAC P-40s are caught on the ground as the first Japanese bombs explode during the attack on Pearl Harbor. Relax, it's just a realistic recreation of the Day of Infamy courtesy of the ordnance experts of the Confederate Air Force. [*Mike Jerram*]

ABOVE, LEFT AND ABOVE LEFT Mike DeMarino in Bob Pond's Curtiss TP-40N Warhawk N999CD over the mountains near Chino, California in March 1986. The TP-40N was a dual control variant of the Warhawk with a full instrument panel and set of controls in the rear cockpit and used as an advanced fighter trainer by the Army. During the late 1950s and early 1960s an attempt was made to make the Warhawk look like a single-seater, the complex dual sliding canopies being removed. On exhibition for a considerable period of time at the USAF Museum in Dayton, Ohio, it was declared surplus when a more original P-40E became available and Bob Pond eventually purchased the machine. Seen on a test flight after restoration by Fighter Rebuilders, the aircraft now has its rear seat controls installed. The rugged Curtiss excelled in the ground attack role and did particular damage to the German war machine in North Africa. [*Michael O'Leary*]

P-40 WARHAWK

BELOW During the 1960s aircraft buff Mike Dillon became one of the first to catch 'warbird fever'. Dillon purchased an abandoned P-40N on an airstrip in Texas and put blood, sweat, tears and money into getting the corroding pile of metal flying again. He was successful in his endeavors, but due to an expanding family Dillon had to sell his prized N1226N to the Confederate Air Force in Harlingen. Here astronaut Joe Engle is seen flying the Warhawk. [*Michael O'Leary*]

BOTTOM A particularly significant Kittyhawk, this P-40E N41JA has been restored in the markings of the late Robert Prescott, an ace with the American Volunteer Group who also, perhaps more importantly, started Flying Tiger Lines, the first all-cargo airline, which has developed into one of the world's largest operators. N41JA is owned by the parent company and the tail surfaces of the Kittyhawk (painted as a P-40B Tomahawk) have been signed by surviving members of the AVG. [*Michael O'Leary*]

BELOW AND BOTTOM The P-40 achieved immortality through the exploits of Major General Claire Lee Chennault's American Volunteer Group, 'The Flying Tigers'. The three squadrons which defended Chinese airfields and supply lines between December 1941 and July 1942 racked up 286 kills against Japanese fighters and bombers in aerial combat and a further 240 on the ground for the loss of only 23 US pilots. Since the Warhawk could never match the Zero-Sen's speed, agility and ceiling, Chennault perfected unorthodox hit-and-run tactics, capitalizing on the aircraft's ruggedness, weight and superior diving qualities and the fighting prowess of his pilots. Though often characterized as a bunch of unruly mavericks – the mercenaries were each paid $700 a month for their services – the AVG pilots honed themselves into a fine fighting force which used superior tactics to overcome the P-40's inferiority to Japanese fighters and strike significant blows, often with as few as 20 airworthy aircraft. [*Mike Jerram*]

ABOVE, LEFT AND ABOVE LEFT 'Trust a woman to pick a color like that!' observed a bystander when Suzanne Parish of Kalamazoo, Michigan arrived at Oshkosh in her boudoir pink Curtiss P-40N. Not fair, sir. Though perhaps a trifle garish to the purist, the scheme faithfully represents the desert sand shade applied to USAAF aircraft during the North African campaign which bleached and weathered to a dusty pink hue.

The P-40N Warhawk was the final production version of the Curtiss fighter. With a 1200-hp Allison V-1710-81 engine it had a maximum speed of 343 mph and carried six .50 machine-guns. The last aircraft left the assembly line in December 1944 after 13,738 P-40s had been built.

Though inferior to most of its contemporaries, the P-40 was the only 'modern' fighter in the American inventory when the Unites States went to war with the Japanese, yet it remained in production until the end of 1944, long after more advanced types such as the P-51 Mustang were available. Why? Ability to absorb punishment and still come home was one reason, eloquently encapsulated by one P-40 pilot who declared that his plane came back from missions 'so full of holes you had to put it against a dark background to see it'. [*Mike Jerram*]

P-40 WARHAWK

BELOW The pugnacious nose profile of the Curtiss P-40 inspired the much-imitated shark-mouth decoration featured on so many American fighters during World War 2. This fearsome looking Warhawk belongs to the legendary Confederate Air

BELOW How long can you get – and what's the propeller clearance?! Ray Hanna knows – that's why he flies The Old Flying Machine Company's Curtiss P-40 Kittyhawk with the touch of the master. [*Norman Pealing*]

BOTTOM Completely rebuilt from a hulk, this P-40E Kittyhawk, painted up in immediate prewar US Army Air Corps markings, is the property of P-40 enthusiast Dr Bill Anderson. Almost completely original, N940AK is a fine example of loving warbird restoration. [*Michael O'Leary*]

P-40 WARHAWK

BELOW During World War 2 airmen and ground-crews produced something which is only now seen as a very particular form of folk art: the aircraft pin-up. *Sneak Attack* adorns the tail of John Paul's Curtiss P-40E Kittyhawk, currently operated in Britain by Ray and Mark Hanna. [*Michael O'Leary*]

BELOW Fearsome teeth highlight the menacing jaws of John Paul's P-40E. [*Michael O'Leary*]

ABOVE, LEFT AND FAR LEFT This carefully researched paint scheme is the result of years of work by aircraft restorer Eric Mingledorff. Eric purchased P-40E (an ex-Royal Canadian Air Force fighter, like most surviving P-40s), and spent several years restoring N1207V to absolutely stock condition. A stickler for detail, he picked the rather distinctive paint scheme used by Lt Dallas Clinger and not only traced the pilot but also his crew chief to obtain the precise details of 14th Air Force colors and markings. Clinger still has the rudder fabric from his aircraft and Mingledorff was able to create an exact reproduction. After performing at airshows for several years he reluctantly sold his beautiful Kittyhawk and went on to concentrate on breathing life back into the neglected airframe of an FM-2 Wildcat.
[*Michael O'Leary*]

P-40 WARHAWK

Bare bird: the Air Museum at Chino, California needed a flying P-40 to complement their collection of rare gunfighters so NL45104 was built up using a bare airframe and parts acquired from many sources. After restoration, the P-40N was flown for several years in a bare metal finish with minimal national insignia and a hungry set of teeth painted on the cowling. Steve Hinton is seen here piloting the Warhawk during the 1983 Minter Field Air Museum Warbird show. [*Michael O'Leary*]

ABOVE A late-model P-40N Warhawk, the Curtiss proved to be a welcome addition to The Air Museum's growing fleet of gunfighters and, since completion, has been used in several movies and television programs. The N-model Warhawk was the ultimate P-40 built by Curtiss, a total of 5219 rolling off the production line. Lightened considerably during the design phase, the P-40N was the fastest member of the Warhawk family, its top speed being about 378 mph. [*Michael O'Leary*]

P-40 WARHAWK

BELOW Hawk over the barren hills of Bakersfield, California. Not so many years ago the Curtiss P-40 was a distinctly rare shape in the skies over America, only a couple of examples being airworthy. With the increased interest in warbirds, basket-case Curtiss fighters have been rebuilt to concourse condition but restorer Bill Destefani was extremely lucky – he found a complete P-40M in an auto museum at Reno, Nevada. [*Michael O'Leary*]

BOTTOM A farmer by trade, Destefani bases his P-40M and racing Mustangs out of Minter Field, Shafter, California – an old World War 2 RAF and USAAF primary training base. [*Michael O'Leary*]

RIGHT AND BELOW Storing his rare find for several years while he completed a couple of Mustang projects, Destefani was able to lavish considerable attention on the Curtiss when the restoration started, virtually every nut and bolt in the fighter being new. A modernized cockpit and instrument panel was installed and the completed Kittyhawk, registered N1232N (which had been used as a cloud seeder in rain-making experiments during the 1950s and early 1960s), was finished in a pristine Royal Air Force paint scheme. [*Michael O'Leary*]

Used in large numbers by the US Army Air Force and the Soviet Air Force, the unconventional Bell P-39 Airacobra was unique in having a nosewheel-type landing gear and the 1325-hp Allison V-1710-63 vee-12 behind the pilot. The propeller driveshaft ran under the pilot's seat and the reduction gearbox was located in the nose. Most P-39s were armed with a 37 mm cannon firing through the propeller hub, two .50 caliber machine-guns in the nose synchronized to fire past the propeller and two .30 caliber machine-guns in the outer wings. The P-39 was the favored mount of many Soviet aces. [*Norman Pealing*]

BELOW Few warbird types are rarer than the attractive mid-engine fighters produced by the Bell Aircraft Company. The P-39 Airacobra and P-63 Kingcobra offered the promise of great performance but failed to deliver – due mainly to the increasing Army demands placed on the airframes, detracting from the original interceptor mission. Today, only two Airacobras and three Kingcobras are airworthy and one of the most familiar to airshow spectators is NL62822, a Kingcobra currently owned by Bob Rieser. Built up during the 1970s as a high-performance unlimited air racer by John Sandberg, the much-modified P-63 never did really well against the Mustangs and Bearcats but it was still a strong competitor. Passing through several owners since then, the P-63 has acquired a few of its more original features, such as wing tips, but the airplane is so highly modified that it is only representative of the type in general outline. Retaining the bright red paint of its racing days, the Kingcobra is similar to some of the 'pinball' P-63s, heavily armored aircraft used as flying targets for student gunners and usually painted in bright schemes for obvious reasons.
[*Michael O'Leary*]

RIGHT This unique Bell P-63F N6763 has survived two unlimited racing careers and a number of private owners. Currently owned and operated by the Confederate Air Force, the P-63F is seen over Texas in May 1986. The F variant of the Kingcobra was a one-off attempt by Bell to update and improve the basic design. The airframe was generally cleaned up to offer better streamlining and a much larger vertical tail was added to improve stability but, with several more advanced designs to choose from, the Army Air Force didn't proceed with the P-63F – and N6763 became the last of Bell's propeller-driven fighters.
[*Michael O'Leary*]

Sea fighters: BEARCAT, HELLCAT & WILDCAT

The Grumman 'Cat' family of aircraft is one of the largest and most successful groups in military aviation history. Starting effectively in the early 1930s with the FF-1, Grumman has supplied fighters for carrier decks ever since, the company transitioning successfully from biplanes to monoplanes, piston engines to jet engines and straight-through wooden decks to acres of angled steel.

The classic litter of the 1940s began with the portly Wildcat, the first 'modern' fighter for the US Navy and an aircraft which slowly enabled the Americans to claw their way back into the Pacific. The unassuming Hellcat soon followed, an aircraft that was a firm favorite with its crews and the mount of many leading Navy aces. Finally came the stocky Bearcat, an aircraft built like a champion prize-fighter and capable of delivering a fast knockout punch. Pugnacious beauty is the common denominator among the 'Cats', the large rasping Pratt & Whitney up front giving the ferocious felines both distinct sound and appearance.

LEFT The late Stefan Karwowski presents the no-nonsense visage of Stephen Grey's Grumman F8F Bearcat during a hop from the fighter's base at Duxford, Cambridgeshire, England.
[*Norman Pealing*]

LEFT Hugh Proudfoot starts the R-2800 of Stephen Grey's F8F-2P, the one Bearcat currently flying in Britain. NX700H (BuNo 121714) is a regular headturner at airshows in both the States (*below*) and Europe. First flown on 21 August 1944 (XF8F-1), the Bearcat was powered by a Pratt & Whitney R-2800-34W Double Wasp 18-cylinder radial rated at 2400 hp with water injection for take-off. [*Michael O'Leary*]

RIGHT Grey's cowling boasts the insignia of VF-11 'Red Rippers'. [*Michael O'Leary*]

BELOW Grumman's F8F Bearcat was the last of the great 'cat' single-engined fighters. This F8F-2 carries the markings of VF-11 *Red Rippers*, one of the oldest US Navy fighter squadrons and now flying F-14A Tomcats. VF-11's insignia on the cowling depicts a wild boar's head (taken from a gin bottle label), some baloney, two red balls and a lightning bolt. Students of heraldry will interpret this as meaning gin-drunken, baloney-slinging, fast-moving bastards . . . Although it was photographed in the Unites States, this Bearcat is now based in Europe with British collector Stephen Grey. [*Mike Jerram*]

ABOVE A close-up of Howard Pardue's rare pre-production XF8F-1. As the Hellcat was a logical extension of the Wildcat, the Grumman Bearcat was a similar extension of the F6F. The Navy wanted the 'smallest airplane with the biggest engine'. Selecting the tried and true R-2800, the F8F Bearcat combined an elegant, streamlined airframe with all the available aeronautical technology that had been developed from combat experience. The Bearcat gained the distinction of being one of the finest propeller-driven aircraft ever built – and still holds the piston-engine time-to-climb record. [*Michael O'Leary*]

BELOW Pardue's magnificent machine in flight. It's interesting to compare the Bearcat with the Corsair (designed and flown before World War 2); both are powered by the R-2800 but the Bearcat reflects the sophistication gained in aeronautical design during wartime. A superlative carrier-based fighter-bomber or night fighter, the Bearcat was agile, extremely fast (the F8F-2 had a top speed of 447 mph) and, at 5000 feet per minute almost straight off the deck, climbed like the original homesick angel. [*Michael O'Leary*]

LEFT AND BELOW LEFT The Bearcat was a successful attempt to combine massive power with a small airframe to provide climb and roll rates twice those of its predecessor the F6F Hellcat. The Double Wasp engine gave this hotrod a climb rate of 5600 feet per minute and a maximum speed in later versions of nearly 450 mph. One novel feature of the F8F-1 was its 'safety' wing tips, boasting expendable outer panels which were supposed to shear off by means of explosive bolts if the wing was overstressed by excessive G loads during high-speed manoeuvring. But asymmetric departures of wing tips resulting in roll rates even more spectacular than usual killed that enterprising idea. The Bearcat pictured here is an immaculately restored F8F-2 wearing the colors of the Naval Reserve Air Center, Denver, Colorado. [*Mike Jerram*]

BELOW This Bearcat is unique: it was built as a civilian aircraft and never saw military service. N700A (designated a G-58B and not an F8F), was used as a high-speed transport by the company's field service to call on Navy and Marine units equipped with Grumman jets. [*Michael O'Leary*]

RIGHT Hinton pictured against the background of the mountains near Chino, California, in 1986. Bearcats can still be seen rounding the pylons at the Reno races too, although they seem to lack the competitive edge seen in earlier years. In August 1969 Darryl Greenamyer set a new mark for non-jet aircraft by squeezing 482.5 mph out of a specially modified Bearcat. The P-51D *Dago Red* raised the record to 517.06 mph on 30 July 1983. [*Michael O'Leary*]

BELOW The 'non-fighting' cat, purchased by Bob Pond in 1987, is exercised by Steve Hinton. The Bearcat arrived on the scene too late to see active service, but it was used for ground-attack missions by the French in Indo-China. [*Michael O'Leary*]

F8 BEARCAT

BELOW Grumman's logical follow-on to the Wildcat was the very capable F6F Hellcat, the US carrier-borne fighter that finally laid to rest the legendary invincibility of the Japanese 'Zero'. Powered by the reliable R-2800 (*the* US Navy engine of World War 2), the Hellcat could take on anything the Japanese could put in the air. Hellcats served with Reserve units after the war yet today only about ten remain flyable. One of the best of recent restorations is the F6F-5 belonging to Bob Pond. Brought back to life by Steve Hinton's Fighter Rebuilders, N4964W is seen here after emerging from the paint shop in a tri-color camouflage scheme. [*Michael O'Leary*]

BELOW Complete restoration: Pond's Hellcat in
formation with The Air Museum's own F6F-5, flown
by Mike DeMarino. Postwar the Hellcat rapidly
disappeared from Fleet squadrons and many
airframes were consumed during the course of
air-to-air guided missile trials. [*Michael O'Leary*]

ABOVE Ex-Grumman test pilot Dick Foote pictured aloft in a dark blue Wildcat sporting USS *Tulagi* escort carrier markings. N11FE is of interest because its rotund fuselage has been modified to hold four passengers – the 'wide body' fuselage being more than adequate for the task! Looking closely, you can see the smoked windows for the passengers. [*Michael O'Leary*]

LEFT Profile of a 'fat cat'. The Wildcat was conceived as a biplane but emerged as the US Navy's first modern fighter, living up to its name to fight magnificently against the Japanese onslaught in the Pacific until relieved by the F6F Hellcat at the end of 1943. Wildcats continued to operate from the smaller escort carriers until the end of the war, including the heroic defence of Wake Island. [*Michael O'Leary*]

BELOW One of the better-known Wildcats, Howard Pardue's FM-2P is a rare photographic variant: note the camera port underneath the wing root. The rather complex undercarriage arrangement of the Wildcat was extremely sturdy and ideally suited to carrier ops, but the manual deployment and retraction mechanism that came with it was not overpopular with the pilots, 29 turns of the hand crank being needed to retract the gear. Needless to say, you could tell a Wildcat pilot by his bulging biceps. [*Michael O'Leary*]

F4 WILDCAT

Howard Pardue's and the CAF's FM-2 form an attractive duo over a Texas cloudscape. Both machines are often used in Confederate Air Force shows to repel Japanese forces (represented by modified AT-6 and BT-13 trainers used in the film *Tora! Tora! Tora!*). [*Michael O'Leary*]

Above N681S, the Wildcat owned and operated by the Confederate Air Force. Impeccably rebuilt, the plane is finished in the very attractive prewar markings of an aircraft serving aboard USS *Ranger*. All flying Wildcats are Eastern-built FM-2s. [*Michael O'Leary*]

Left and Right The Grumman F4F/FM Wildcat (or Martlet in British Fleet Air Arm service) was a marvelous naval fighter – feisty, solid, dependable and manoeuvrable. [*Norman Pealing*]

BELOW Fighting all the way through the war, the Wildcat had little use after the end of hostilities and most were quickly scrapped. However, a few were purchased surplus by private owners and put to work in a variety of menial tasks including crop spraying and aerial survey. N315E is owned by Les Dupont of Wilmington, Delaware. [*Michael O'Leary*]

BELOW In recent years over a half dozen Wildcats have been discovered and restored, giving a flying population of about a dozen machines. Bob Pond's N47201 is seen at the Imperial War Museum at Duxford after being painted in a rather fanciful interpretation of the markings of US Navy Commander Butch O'Hare's Wildcat. [*Michael O'Leary*]

BOTTOM Awkward, narrow-track 'roller skate' landing gear became a Grumman hallmark after its introduction on the 1931 FF-1 biplane fighter, but was an invention of the Loening company, for which Grumman founder Leroy Grumman, Leon Swirbul and Bill Schwendler had worked. This aeroplane, owned by Joe Frasca, is a late production FM-2 model built by General Motors. The landing gear was hand-cranked, 29 turns to retract it, and was a sure give-away for a newcomer to the type because the cranking motion would transmit to his stick hand, remitting in an undulating flight path after take-off until the gear was up. [*Mike Jerram*]

Air pirate: CORSAIR

The Vought F4U Corsair was a very different naval fighter to the relatively diminutive Wildcat, being a real 'muscle plane' with tremendous power and a formidable offensive capability. The XF4U-1 prototype made history by becoming the first US warplane to exceed 400 mph in level flight. The aircraft reflects certain design points of the 1930s, including the fabric covering from the wing spar on the back and the wooden ailerons.

RIGHT Deadly duo: Rick Brickert flies Stephen Grey's Goodyear-built FG-1D Corsair in the foreground while Mike DeMarino keeps him company in The Air Museum's FG-1D. Grey's Corsair had just completed an overhaul at Fighter Rebuilders and was on its way to Florida, where it was put on a boat and shipped to Britain for the 1986 airshow. [*Michael O'Leary*]

LEFT Howard Pardue's Vought F4U-5N Corsair NX65HP, a fighter that last saw service with the Honduran Air Force. Basically a radar-equipped night-fighter (the large wing-mounted radar pod has been removed), the -5N was a fast and efficient machine, its four 20 mm cannon proving particularly deadly. [*Michael O'Leary*]

BELOW LEFT The F4U-5N Corsair night-fighter was used throughout the Korean War. Pardue's restored example, lacking only the '5N's radome normally mounted on the starboard wing, wears the colors of *The Flying Nightmares* – Marines squadron VMF (N) 513. Lt Guy Bordelon, an F4U-5N pilot with US Navy squadron VC-3, became the only Navy ace of the Korean War with four Communist YAK-18s and one LA-2 downed. [*Mike Jerram*]

BELOW The Air Museum's FG-1D prepares for a flight. Initially seeing combat with the US Marine Corps (the Vought Corsair had a problem landing aboard the Navy's carriers – the stiff gear made a smooth landing almost impossible and it took some time to work this problem out), the F4Us quickly compiled a distinguished combat record during the island-hopping Pacific war. [*Michael O'Leary*]

BELOW One of the main reasons for the Corsair's success: the famous P&W R-2800. [*Michael O'Leary*]

Howard Pardue's superb FG-1D 'down and dirty', illustrating the imposing nature of the Corsair. Because of the poor view over the long nose, however, the US Navy judged the Corsair to be unsuitable for carrier operations (a view not shared by Britain's Fleet Air Arm), and unusually the Marine Corps accepted all the early deliveries. Around two dozen Corsairs are flying or being restored. [*Michael O'Leary*]

F4U CORSAIR

BELOW Close formation flying at its best: Howard Pardue (trained as a US Marine Corps fighter pilot) shows what formation flying is all about. His FG-1D, NX67HP, was photographed over Chino, California, during May 1983 from the back seat of a Beech T-34 Mentor using a 24 mm lens. [*Michael O'Leary*]

RIGHT *Big Hog* is a Goodyear-built FG-1D which last saw active service with the Royal New Zealand Air Force, an air arm that made extensive use of the 'bent-wing bird' during World War 2. Rescued from a scrapyard, the plane was transported back to the US and eventually completely rebuilt by the late Jim Landry. He incorporated many updates in the Corsair, including metalized wings, extra fuel and

new avionics – everything needed to make the fighter a safe and efficient cross-country machine. Landry was later to die in a skydiving accident. [*Michael O'Leary*]

BELOW RIGHT One of the few drawbacks to the Corsair during its operational service was the rather poor visibility over that long nose during take-off and landing. The pilot, positioned roughly mid-fuselage, had a long chunk of aluminum to look past during those two important phases of flight. Once in the air, the pilot of the Corsair enjoyed pretty good visibility, as can be seen in this banking view of Howard Pardue's F4U-4, NX68HP – another of the Corsairs returned to the States from Honduras in the late 1970s. [*Michael O'Leary*]

F4U CORSAIR

A formation that shows Corsair power at its best: the late Merle Gustafson leads the pack in his F4U-4, followed by Buck Ridley in his F4U-4 and Howard Pardue in his F4U-5N. The Corsair stayed in production longer than any other American piston-engined warplane. [*Michael O'Leary*]

LEFT About to touch down is a Goodyear-built FG-1D Corsair owned by the Canadian Warplane Heritage. CWH president Dennis Bradley bought it, with just 1600 hours on the airframe and 300 on the engine, for $25,000. [*Mike Jerram*]

BELOW LEFT The dedication on CWH FG-1D is to Lt Robert Hampton Gray of the Royal Canadian Volunteer Reserve, who was awarded a posthumous Victoria Cross for pressing home an attack on a Japanese destroyer in the Bay of Onagawa on 9 August 1945 despite repeated flak hits. Gray crashed into the sea as his target sank. [*Mike Jerram*]

BELOW The last combat Corsair: Honduran Air Force mechanics prepare to fire up that country's last Corsair, an F4U-5NL. In 1969 Honduras and Salvador went to war and both sides flew Corsairs, the Salvadorian Air Force beefed up by P-51s smuggled south by American pilots after a quick buck. This aircraft, flown by Fernando Soto, shot down two Salvadorian FG-1Ds and a P-51D on one mission – the last World War 2 piston-engine fighter to record a kill over a similar type of machine. [*Michael O'Leary*]

F4U CORSAIR

BELOW The classic lines of the Corsair are shown to advantage high above a Texas coastline. This FG-1D was rescued from El Salvador. [*Michael O'Leary*]

RIGHT Two FG-1Ds in formation: Mike DeMarino leads in The Air Museum's example (a veteran of the *Baa, Baa, Black Sheep* television series) while Howard Pardue follows in his Goodyear-built Corsair, painted in the markings of VMF-111. Both these airplanes have been modified to carry a passenger directly behind the pilot, and the blue-tinted plexiglass of the passenger's window is just visible. Two-seat modifications are becoming more common on the surviving gunfighters – giving someone else a chance to experience the thrill of this type of flying. [*Michael O'Leary*]

BELOW RIGHT For some curious reason the rather elegant Corsair has collected a wide variety of pig-related nicknames. Here Mike Wright enjoys some summer flying with the canopy slid back in Don Davis's FG-1D *Wart Hog*. N4715C was completely restored by The Tired Iron Racing Team in Casper, Wyoming, and is flown as both a warbird and as an unlimited air racer, though participating in a leisurely manner. This fighter is also equipped with gas-operated 'machine-gun' – units that give off a noise convincingly like .50 caliber Brownings and, needless to say, extremely popular with airshow crowds. [*Michael O'Leary*]

LEFT From the moment the Marines hit the Solomons in February 1943 their Corsairs had the enemy on the run. Demoralized Japanese pilots dubbed the bent-wing bird 'The Whispering Death'. The Marines are a determined bunch, as one Lt Robert Klingman demonstrated during a combat with a Japanese *Nick* fighter at 38,000 feet: finding his guns had frozen, he coolly sawed off the tail surfaces of the opposing fighter with his propeller. [*Norman Pealing*]

BELOW As the beefed up AU-1 the Corsair saw further action in Korea, but by this time the aircraft had lost its fine handling qualities, proving a real handful for the average pilot. The monster F2G was even worse, ploughing behind a 3000-hp Pratt & Whitney R-4360 Wasp Major 28-cylinder four-row radial. Corsair production ended as late as December 1952 at number 12,571. [*Norman Pealing*]

BELOW Chance-Vought's Corsair naval fighter earned the nickname 'bent-winged bird' because of its unusual wing configuration, designed to enable short undercarriage legs to be used while still providing adequate ground/deck clearance for the massive propeller. This fine example belongs to the Confederate Air Force. [*Norman Pealing*]

Stormy weather: THUNDERBOLT & LIGHTNING

The P-47 Thunderbolt was the crowning masterpiece of Republic's chief designer Alexander Kartveli. Mounting eight .50 caliber Browning machine-guns and a variety of underwing stores, it was capable of creating havoc among enemy ground forces. But that big P&W really gulped the fuel and prevented the Thunderbolt (which had nicknames like *Jug*, *T-Bolt*, and *Flying Brick*) from escorting 8th Air Force bombers deep into enemy territory. Additional fuel tanks helped the situation but the Thunderbolt never really became an efficient long-range escort fighter, though it certainly played its part until the Mustang became available.

LEFT Ray Stutsman's magnificent P-47G restoration poses proudly above California's central valley. [*Michael O'Leary*]

P-47 THUNDERBOLT

The P-47's juggernaut proportions were determined
by the massive Pratt & Whitney R-2800 Double Wasp
18-cylinder two-row radial (output grew from
2000 hp in the B model to 2800 hp in the P-47M/N),
the bulky rear fuselage turbo and formidable
eight-gun Browning armament. [*Michael O'Leary*]

CAPT. W. C. BECKHAM
SGT. H. BUSH
SGT. M. EICHSTAEDT

LEFT Ray Stutsman has acquired a reputation for meticulous warbird restorations but the Thunderbolt is perhaps his finest effort. Rescued from the junkyard of a Los Angeles aviation eccentric (where it was rotting with several other rare World War 2 aircraft), the P-47 underwent a ground-up restoration at Stutsman's shop in Indiana. While taking the T-Bolt completely apart, Ray found a number of curious items—including a bag of cardboard gun barrel plugs, installed when the aircraft left the Curtiss factory! [*Michael O'Leary*]

RIGHT Production of the Thunderbolt totaled 15,660, including a staggering 12,603 D models from Republic's Farmingdale – which, thanks to the demise of the Fairchild T-46 Air Force trainer, no longer manufactures aircraft – and Evansville, Long Island plants. It's hard to believe that so many aircraft can simply disappear, but when the Confederate Air Force began searching for P-47s in the late 1960s they failed to find a single example in airworthy condition. [*Michael O'Leary*]

BELOW With gear down the Thunderbolt appears to be even more massive. Painted dark Olive Drab and neutral gray, Stutsman selected the markings of Captain W. C. Beckham's *Little Demon*. [*Michael O'Leary*]

BELOW Few shots could be more expressive of the size of fury of the Republic Thunderbolt than this tight formation being flown by John Maloney in the Curtiss-built P-47F and Steve Hinton in the bubble-top P-47M. During the 1960s, the Thunderbolt was virtually extinct from American skies but an infusion of six Thunderbolts from Peru along with several other complete restorations brightened the situation. Now airshow crowds can enjoy the spectacle of superb aircraft like these in flight. [*Michael O'Leary*]

BELOW Steve Hinton in Thunderbolt NX47DD leads an interesting flight of gunslingers: a bubble-top T-Bolt, razorback P-47G, bubble-top P-51D and razorback P-51B. One of the ex-Peruvian AF machines, NX47DD is now based at Duxford in England in a new paint scheme. [*Michael O'Leary*]

BOTTOM Obtained in the early 1950s from a trade school, Ed Maloney's P-47G had never seen combat and had accumulated few flying hours before being surplused off to teach a new generation of mechanics. [*Michael O'Leary*]

P-47 THUNDERBOLT

BELOW AND BOTTOM The Air Museum's P-47G airborne in the capable hands of Don Lykins. NX3395G (s/n 42-25234) made its first flight after restoration during April 1985 and is painted in the markings of World War 2 and Korean ace Walker 'Bud' Mahurin. A rear seat has been added to give passengers the experience of flying in one of America's truly great combat aircraft. Heavily damaged during a forced landing after engine failure at an airshow in the early 1970s, Ed Maloney's Thunderbolt was put into storage for many years before being completely rebuilt. [*Michael O'Leary*]

BELOW RIGHT *No Guts–No Glory!* The Thunderbolt pulverized the Axis throughout the European, Pacific and Far East combat zones, strafing, bombing, rocketing and proving a surprisingly agile fast-climbing dogfighter. [*Norman Pealing*]

BELOW RIGHT The perfectly streamlined bubble canopy sets of this close up portrait of the biggest, most powerful single-engined fighter to see combat in World War 2. [*Norman Pealing*]

BOTTOM N47DF having its R2800 run up before a 1974 flight. Six P-47s were returned to the United States in the late 1960s after serving with the air force of Peru. Purchased by vintage warplanes collector Ed

Jurist, they were taken to the main Confederate Air Force base at Harlingen, Texas. After being assembled the machines were flown in formation at several airshows – the first time a massed P-47 flight had taken place in the States since the early 1950s. Jurist later sold the group to another collector and several of these rare fighters have, unfortunately, been written off in accidents. [*Michael O'Leary*]

LEFT The Germans called it *Der Gabelschwanz Teufel* – the fork-tailed devil; the French said *Double Queue* – twin-tailed one; the Japanese had a symbol for it meaning two aeroplanes, one pilot; and the British and Americans called it simply Lightning. Whichever you prefer, Lockheed's P-38 was an innovative aeroplane: the first twin-boom fighter, first with tricycle landing gear, first with turbo-supercharged engines, the first twin-engine, single-seat interceptor and the first to make extensive use of stainless steel in its airframe. Lightnings are rare birds indeed: this pristine example is a P-38L owned by Bill Ross of Lake Barrington Shores, Illinois. [*Mike Jerram*]

BELOW LEFT Lockheed P-38M NL3JB, a two-seat night fighter, during one of its vary rare outings from the Champlain Fighter Mueseum in Mesa, Arizona – home of a fabulous collection of World War 1 and World War 2 gunfighters, all of them in flying condition. [*Michael O'Leary*]

ABOVE The P-38 was a product of Lockheed's legendary designer Clarence 'Kelly' Johnson, who later directed the design of the P-80 Shooting Star, F-104 Starfighter, U-2 spyplane and the SR-71 Blackbird. [*Michael O'Leary*]

P-38 LIGHTNING

BELOW An extremely advanced warplane at the time of its first flight on 27 January 1939, the Lightning was Lockheed's bold response to a February 1937 US Army Air Corps specification for a long-range pursuit and escort fighter capable of sustaining 360 mph for one hour at 20,000 feet. [*Michael O'Leary*]

BELOW Although less manoeuvrable than contemporary single-engined fighters, the P-38 endeared itself to pilots by proving to be exceptionally reliable, long-ranged, hard-hitting and fast – up to 414 mph in later versions. When the US declared war on Japan and Germany on 7 December 1941 one hotshot P-38E pilot lost no time in drawing first blood, destroying a snooping FW 200C Condor near Iceland within a few minutes of the announcement.

The Condor stood little chance against the E model's battery of one 20 mm Hispano cannon, four .50 Browning and two Colt .30 caliber machine-guns. This armament was standard on most P-38 models. Another notable kill for the P-38 came when 16 aircraft of the 339th Fighter Squadron shot down Admiral Yamamoto's G4M transport after flying 550 miles from their base on Guadalcanal. [*Mike Jerram*]

RIGHT 'Lefty' Gardner and his P-38 *White Lit'nin* are firm favorites at Harlingen and the Reno air races. Gardner's superbly crafted display is certainly on the 'unmissable' list of every airshow devotee. [*Norman Pealing*]

BELOW The P-38 featured a string of hi-tech items, including a tricycle landing gear, Fowler flaps and the new Allison V-1710 vee-12 glycol-cooled engine with GEC turbos recessed into the tail booms, cooling radiators on the sides of the booms and induction intercoolers in the wing leading edges. The beautifully streamlined central nacelle accommodated the powerful nose armament and pilot. [*Norman Pealing*]

RIGHT AND BELOW After many years' work and the expenditure of a large amount of money, John Siberman brought a badly wrecked Lockheed F-5 (Lightning photo-reconnaissance variant) back to life and his rare machine is pictured airborne during the 1986 Valiant Air Command airshow in Florida. Difficult to restore, with many complex systems, around four Lightnings are currently under rebuild. N5596V is being flown here by Thurston 'Jaybo' Hinyub, accompanied by Don Davidson's Mustang *Double Trouble Two*. The P-51 is pictured in more detail on page 27. [*Michael O'Leary*]

Light fantastic 1: AVENGER

Deceptively large, with a wingspan of 54 ft 2 in and weighing up to 17,895 lb (TBM-3E), the Grumman TBF/BM Avenger was the US Navy's standard carrier torpedo bomber in World War 2. The Avenger had a literal baptism of fire when, on their first mission in the Battle of Midway in June 1942, five aircraft of VT-8 were shot down by Japanese fighters; the sixth machine in the formation returned to the USS *Hornet* as scrap metal with a dead gunner.

Happily, this mission was totally untypical and the Avenger went on to fly supreme for the remainder of the Pacific war, sending (among others) the great super battleship *Musashi* to the bottom. In addition to destroying the Japanese Imperial Fleet, the Avenger and its legendary partner, the SBD Dauntless, dispatched millions of tons of enemy merchant shipping. Grumman built 2290 Avengers, while Eastern added another 2882 TBM-1s and 4664 TBM-3s to the Navy ranks.

LEFT Probably the finest example of a restored Avenger currently flying is Dr John Kelly's TBM-3E, which has been fully outfitted back to its late World War 2 configuration. N9586Z has been furnished with the Avenger's normal armament of one forward-firing .30 caliber machine-gun, one .50 caliber in the turret and one .30 caliber in the ventral position. Up to 1600 lb of weapons could be carried in the bomb-bay. [*Michael O'Leary*]

ABOVE The attractive planform of the Avenger's wing is seen to advantage in this high-angle view of Dr Kelly's N9586Z. After the war, Avengers were modified by the Navy and used for a wide variety of tasks including airborne early warning (equipped with a huge radome under the fuselage) and as a COD transport (carrier onboard delivery) with a seven-seat interior. [*Michael O'Leary*]

LEFT Much detailed work brought Dr Kelly's Avenger N9586Z back to such fine condition. When Avenger restorations started in the 1970s most fire bomber bases still had some original parts left in their junkyards – items like turrets and bomb-bay doors that, while having no use to the fire-bombing role, where absolutely essential to restorers. These items have now become extremely difficult to find and in a recent auction a pair of very well-used bomb-bay doors went for a stunning $11,000. [*Michael O'Leary*]

ABOVE High over Breckenridge, Texas, Howard Pardue is seen airborne in TBM-3E NX88H, named *Turkeycat*. Restored to 1943 combat configuration, Howard reports that this particular aircraft was used by the military at one stage in its career for testing ejection seats. This view shows the long bomb-bay to advantage, its length needed to accommodate the single torpedo the aircraft was designed to carry – as the premier US Navy torpedo bomber of World War 2. [*Michael O'Leary*]

TBM AVENGER

LEFT The Avenger is interesting in the fact that it appears portly yet rather sleek at the same time. This contradiction is probably due to the fat fuselage, which housed three crew members, a gun turret and the torpedo. The wing of the Avenger is large but has a rather elegant sweepback to the outer panels, which helps dispel the heavy lines of the fuselage. [*Michael O'Leary*]

ABOVE TBM-3 N66475 was rather typical of the hard-working Avengers used during the 1960s, before the type acquired warbird status and, hence, increased value. This plane has been extensively modified as a sprayer: note the large tank placed in the bomb-bay and the spray booms under the wings. The canopy has been cut down to a single-place configuration. The scruffy condition is indicative of the plane's hard-working role. N66475 is not carried on today's civil register – which means the plane was either crashed, scrapped or sold in Canada. [*Michael O'Leary*]

BELOW Painted Glossy Sea Blue, Avenger NL7001C was restored by Ralph Ponte of Grass Valley, California, for warbird collector Gordon Plaskett. [*Norman Pealing*]

ABOVE With a top speed barely above 260 mph the Avenger is obviously no speedster, but with a reliable 1700 hp Wright R-2600 Cyclone 14-cylinder radial up front and solid, predictable handling, there's little doubt that the Grumman TBF was one of the hardest working combat aircraft of World War 2. There's also little doubt that surplus Avengers were some of the hardest working ex-military aircraft, performing a wide variety of workaday tasks including fire bombing and bug and crop spraying. In California particularly, Avengers were busy fighting the deadly forest and brush fires which attempt to consume the state every summer. However, a ban was finally instigated in the late 1960s on operating single-engine aircraft for Forest Service work. Most of the Avenger fleet was sold to Canada (average price was around $5000), where they went on to do battle with the spruce bud worm infestation that causes extreme damage to that nation's valuable lumber resources.
[*Norman Pealing*]

ABOVE The Avenger enjoyed a long and productive career with the US Navy and Marines. After the final example rolled off the production line in September 1945, many were rebuilt to serve in a variety of roles, including anti-submarine warfare (ASW) as the TBM-3E; airborne early warning (AEW) as the TBM-3W and -3W2; target-towing tug as the TEM-3U; carrier on-board delivery (COD) as the TBM-3R; and TMB-3Q ECM platform. Special night attack versions (TBM-3E and 3N) were also produced. [*Norman Pealing*]

RIGHT Gordon Plaskett's Avengers, here tucked up for the night, stayed in Fleet service until 1954—longer in the Reserves. [*Michael O'Leary*]

TBM AVENGER

BELOW Although blanked off, this view shows where the ventral .30 caliber weapon was installed immediately forward of the tail wheel. The Avenger is just one of Bob Pond's fleet of impressively restored World War 2 aircraft. Britain obtained 921 Avengers through Lend-Lease and some were operated well into the 1950s. [*Michael O'Leary*]

ABOVE Painted in attractive Fleet Air Arm D-Day invasion markings, Bob Pond's Avenger is seen with its gear down during a May 1983 photo sortie.

TBM-3E NL7075C, BuNo 53785, was operated as a sprayer by Charles Reeder of Twin Falls, Idaho, before being sold for restoration. [*Michael O'Leary*]

BELOW Painted in dark blue camouflage, TBM-3E C-CWG is owned and operated by the Canadian Warplane Heritage and seen here during a moment of rest at the museum's base at Hamilton, Ontario. Obtained from Ed Maloney during the 1970s, the Avenger was a significant addition to the museum since the TBM played a major role with the Royal Canadian Navy, and some of today's band of surviving Avengers are former RCN machines that were sold as surplus by them in the United States. [*Michael O'Leary*]

BOTTOM Grumman's unique Sto-Wing mechanism put the aircraft's wings flush alongside the fuselage for maximum space saving in cramped aircraft carrier hangars and deck parking spots. Here Dick Dieter's General Motors-built TBM-3R Avenger torpedo bomber does a fine impression of a pelican tucking up its wings. Huge greenhouse canopy and portly fuselage stripped of military hardware enables Dieter, from South Bend, Indiana, to be generous with joyrides for friends, as many as seven at a time. [*Mike Jerram*]

Light fantastic 2: SKYRAIDER

Call it *Able Dog*, *Spad* or what you will, Douglas's AD Skyraider attack aircraft is a big brute of an aeroplane which someone once said could carry everything but the kitchen sink. Pilots from US Navy attack squadron VA-195 proved that wrong when they raided the ablutions aboard USS *Princeton*, slung a cast-iron sink from a 2000 lb bomb and dropped the lot on a Communist position during the Korean War.

The Skyraider's basic design was drafted by Ed Heinemann, Leo Devlin and Gene Root during a brainstorming session in a room at the Statler Hotel in Washington one 1944 night to meet a Navy Bureau deadline for contract submission to operate a new single-seat BT category of attack aircraft. It was too late for World War 2 but served with distinction in attack, dive-bomber and early warning roles with the Navy in Korea and was resurrected for use in Vietnam because of its prodigious weapons carrying capability and ability to absorb groundfire while remaining in the air – something sophisticated modern jets could seldom manage.

LEFT George Baker flying Harry Doan's 'battle'-damaged AD-4NA NX91945 over Florida during March 1987. This aircraft was damaged in a landing accident when owned by Jack Spanich (see pages 144–145) and has only recently been returned to flying status. Note the missing landing gear fairings and replacement left wing. [*Michael O'Leary*]

AD SKYRAIDER

Heavy-hitters in formation: the March 1987 edition of the popular Valiant Air Command warbird show in Florida hosted these two former *Armee de L'Air* AD-4NA Skyraiders. George Baker leads the formation in NX91945 (second use of registration) while Dr Bill Harrison flies wing in NX91935.

There's little doubt that during its military service some of the most colorful markings ever carried by the Douglas Skyraider were those of the United States Navy VA-176. The 'Thunderbolts' insignia comprised a very angry insect thrusting its stinger along the vertical fin – an appropriate marking for the Navy's heavy-hitting Skyraider.

The majority of surviving flyable Skyraiders in the Unites States are AD-4NAs imported from France in the 1970s by warbird collector Jack Spanich. When

trouble in Algeria picked up, the French found that they didn't have a really efficient aircraft with which to bomb their anti-colonists. With the immediate aid of the US government, 93 Skyraiders were quickly transfered from Navy stocks to the *Armee de l'Air*. The group comprised 40 AD-4NAs and 53 AD-4Ns which were brought up to AD-4NA standards on arrival in France.

After serving France effectively against the *Front de Liberation Nationale* the fleet of Skyraiders was handed over to Chad, the Central African Republic, French Somaliland, Madagascar and Cambodia following the granting of independence to Algeria. Of the small number of flyable Skyraiders, three are painted in the markings of VA-176.
[*Michael O'Leary*]

AD SKYRAIDER

BOTTOM Spanning over 50 feet, the Skyraider is no small aircraft. Capable of carrying a huge variety of underwing weapons, it was also armed with four 20 mm cannon in the wing. Powered by the oil-loving Wright R-3350, the AD-4N is capable of a top speed of 320 mph. This view of the superb AD-4NA N409Z over Madeira in 1986 gives a good idea of the size of the Skyraider's huge four-blade Aeroproducts propeller. [*Michael O'Leary*]

BELOW AND RIGHT This beautifully restored Spad is actually a former US Navy AD-4N ECM aircraft, masquerading as an A-1H of VA-176, the 'Thunderbolts'. During the Vietnam War an A-1H from VA-176 operating off the USS *Intrepid* and flown by Lt Jg W. Thomas Patton and Lt Peter Russell shot down a North Vietnamese MiG-17 jet-fighter. [*Mike Jerram*]

AD SKYRAIDER

Douglas AD-4NA NX91945 (BuNo 126882) was beautifully restored in Vietnam-era markings by Jack Spanich, one of four AD-4NAs that he returned to the States from France. Photographed over Hamilton, Canada during the annual Canadian Warplane Heritage show in June 1984, Spanich had equipped the Skyraider with underwing rockets, low drag bombs and a huge centerline fuel tank. The AD-4NA was a modification of the -4N which saw all the night attack equipment removed in order for heavier bomb-loads to be carried for Korean operations.
Unfortunately, Spanich and his wife died in NX91945

when the plane slammed into a mountain on 4
November 1984 near Culpepper, Virginia while
Spanich was trying to fly low to avoid bad weather.
The restored aircraft is featured on the three previous
'spreads' of this book in the markings of VA-176
'Thunderbolts'. [*Michael O'Leary*]

LEFT The EA-1E looks a bit awkward with its gear coming down – somewhat in the style of the Curtiss P-40 retraction system. The EA-1E has a large 'hump-back' that originally accommodated four seats (including pilot) for the craft's EW role. During the 1962 Tri-Service designation consolidation, the AD-5W became the EA-1E. [*Michael O'Leary*]

BELOW LEFT The massive lines of the Skyraider are portrayed well in this close-up view of Pete Thelen's EA-1E, photographed during March 1984. Stated by some to be the last Skyraider in operational service with the US Navy, the EA-1E went through a couple of civilian owners, rarely flying, before being purchased by Thelen. N62446 (BuNo 135178) started life as an AD-5W early warning aircraft with a huge under-fuselage radar dome, an almost two-feet

fuselage stretch and a 50 per cent larger vertical tail. Flown for the first time on 17 August 1951, the AD-5 was developed into a number of different variants including the AD-5N night-attack variant, AD-5Q ECM platform and AD-5W early warning variant. Douglas built 218 AD-5Ws. [*Michael O'Leary*]

BELOW During the 1960s most surviving EA-1Es were heavily modified with the removal of the ugly radome and modification of the crew housing, usually having blue-tinted throw-over hatches added for better visibility. The USAF procured 150 A-1Es and had them extensively modified for the Air Commando role in Southeast Asia. The Royal Navy also received 50 AD-4Ws, beginning in November 1951. Today, N62446 is owned and operated by the Lone Star Aviation Museum in Houston. [*Michael O'Leary*]

Medium marvel 1: MITCHELL

Perhaps the best medium/light bomber to serve with any force during World War 2, the B-25 Mitchell saw extensive service in every theater of conflict. Designed by a company with no previous experience in this field, North American's aircraft served for over 20 years basically unchanged. Whether ship-busting in the bays and lagoons across the Pacific, delivering bombs with pin-point accuracy in Northern Italy or routing columns of German Panzers in occupied France, the Mitchell performed with a rugged dependability.

Named after the US Army's controversial and court-martialed champion of air power General William 'Billy' Mitchell, the North American B-25 was immortalized by Jimmy Doolittle's April 1942 raid on Tokyo and other cities, when Mitchells attacked the Japanese after taking off from the aircraft carrier USS *Hornet.*

LEFT *Mitch the Witch* head-on over the mountains north of Chino, California. Photographed from the back of another Mitchell with the gunner's armor glass removed, Bob Pond's freshly restored B-25J has been redone by Steve Hinton's Fighter Rebuilders. The aircraft is painted in the colorful markings of an actual Mitchell of the 17th Tactical Reconnaissance Bomb Squadron that, on 25 February 1944, took on a Japanese *Sally* medium bomber in a bomber version of a dogfight near Kavieng, New Ireland. The Mitchell won. [*Michael O'Leary*]

BELOW, INSET The gaily painted vertical tail of *Mitch the Witch* illustrating the insignia of the 17th Tactical Recon Bomb Squadron (known as the 'Fighting 17th'). Bob Pond's Mitchell divides its time between his museum in Minnesota and Southern California, where it can often be seen at Chino and Palm Springs. [*Michael O'Leary*]

BELOW 'Mitchell, break right now!' *Mitch the Witch* thunders away from the camera plane at the

conclusion of the May 1986 photo-flight. This view shows a number of details to advantage, including the bomb-bay doors, dual landing lights and the tail 'bumper' at the rear of the fuselage. [*Michael O'Leary*]

BELOW From being either workhorses or unwanted eyesores not that many years ago, today's warbird bombers are revered veterans of history's most tumultuous conflict. Accordingly, owners are going to great lengths to make sure their charges are maintained in pristine condition, both mechanically and aesthetically. Paint jobs for the big bombers cost thousands of dollars and are certainly of much better quality than when the aircraft were made. This detail view shows some of the work that went into The Lone Star Aviation Museum's N333RW, including a very nice Varga girl applied ahead of the two .50 caliber Browning 'package' guns. Alberto Vargas (note the 's', which was quite often dropped from his artwork) painted attractive women for the pages of *Esquire* and other publications that became the inspiration for a generation of fighting men. The fact that copies of his seductive females adorned countless combat aircraft was 'one of my proudest achievements' according to Vargas in an interview with this photographer shortly before the artist's death. Nose art was a popular morale booster among USAAF crews in World War 2, and a good squadron artist could earn $15 a time for painting curvaceous features. When they got a little too revealing the top brass tried to impose censorship – after all, what *would* the folks back home think when they saw the publicity pictures? – but cunning artists could soon apply a swimsuit or bikini . . . in washable paint, naturally. Sooner or later it would always rain in Britain. [*Michael O'Leary*]

ABOVE *Dagger dagger dagger!* Has the enemy drawn a bead of this taxiing B-25? Not in an Avenger, we hope: it's merely a spirited fly-by from a navalized Confederate. [*Norman Pealing*]

LEFT Painted in the US Navy's version of Olive Drab, N333RW rumbles down the taxiway at Breckenridge, Texas, during the field's annual May airshow during 1986. Painted as a USN PBJ-8J, the ultra-clean Mitchell well represents the USN and US Marine Corps' heavy usage of the Mitchell in the South Pacific during World War 2. This Mitchell is owned and operated by the Lone Star Aviation Museum, Houston. [*Michael O'Leary*]

B-25 MITCHELL

BELOW Pilot John H. Bell II brings Mitchell
Heavenly Body N8195H (TB-25N, s/n 44-30748, c/n
108-35073) in for a close look at the camera plane.
Owned by Mike Pupich and based at Van Nuys,
California, this Mitchell is an ex-*Catch-22* film star
and has been the subject of much loving care lavished
by a band of volunteers. The large metal patch on the
top fuselage behind the cockpit is a covering over
where the top gun turret was installed. During the
1950s, most surviving military Mitchells had the
turret and its mounting ring removed since the
bomber no longer had a combat role and was used
mainly for transport and training duties.
Unfortunately, the mounting ring and turret are now
extremely difficult (and expensive) to find. The Eagle
Field decal denotes the fact that Pupich and other
warbird lovers have developed a World War 2 civilian
pilot training (CPT) airport known as Eagle Field in
California into an all-warbird base. [*Michael O'Leary*]

ABOVE The twin-finned tail, decorated here with an American Indian chief, is a distinct recognition feature of the B-25. [*Norman Pealing*]

TOP B-25s lined up at Rebel Field before the exciting re-enactment of Lieutenant Colonel Jimmy Doolittle's heroic 18 April 1942 raid against Tokyo and other Japanese cities from the carrier USS *Hornet*. Amazingly, all 16 Mitchells made successful free take-offs from the carrier at a maximum gross weight. [*Norman Pealing*]

B-25 MITCHELL

BELOW *BIG OLE BREW 'n little ole you* is an appropriate name for this restored Mitchell. The B-25's controls are operated purely by 'muscle' power and it can be a tiring aircraft to fly, especially in formation or in poor weather. Beautifully framed by a Texas thunderhead at the September 1986 Wings Over Houston airshow, N5865V is a B-25J, s/n 44-30988, c/n 108-34263 restored by Mitchell

ABOVE N5865V cruises over Texas. Most Mitchell restorers favor the glass nose since an extra passenger can be carried. The B-25J was the final production variant of the Mitchell and 4318 were built (plus 72 further examples scrapped on the production line at the end of the war). The B-25J could carry 3200 lb of bombs and boast 13 .50 caliber air-cooled Browning machine-guns. [*Michael O'Leary*]

RIGHT One of many B-25s at Rebel Field, *BIG OLE BREW* taxies past a galaxy of C-47s, an Airacobra and Kingcobra, a Bearcat, P-51s, a 'Messerschmitt Bf 109' and more! The J model (the most numerous version, accounting for 4318 of 9816 built), is powered by two Wright R-2600-29 Cyclone 14-cylinder radials with an original wartime emergency rating of 1850 hp. [*Norman Pealing*]

rebuilder Tom Reilly, who specializes in bringing basket-case B-25s back to life. Today's detailed restorations are seeing the addition of such items as operable upper gun turrets. [*Michael O'Leary*]

B-25 MITCHELL

BELOW Glowing with the golden light of a setting California sun, Bruce Guberman holds *Executive Sweet* in formation with Ascher Wards's Texan camera plane. N30801 (ex-N3699G) is a B-25J, AAF s/n 44-30801, c/n 108-34076 that led a varied life after being surplused from USAF service. After a bout of spraying bugs the Mitchell became the lead B-25 for the epic film *Catch-22*. When the film was completed the Mitchells were sold off for bargain-basement prices (remember, this was in the early 1970s when few individuals had much use for a fuel-hungry twin-engine bomber). The 'star' aircraft in the movie *Executive Sweet* was purchased by publisher and veteran pilot Ed Schnepf, and he set about to restore the bomber back to its original World War 2 condition. Though the aircraft was flyable, its interior had been gutted of all military equipment and considerable repair and refurbishment was needed to make it a reliable flyer. Several years of hard work went into the bomber and junkyards around the world were searched to find vital military bits and pieces before it was brought back to pristine warbird

condition as the first of a whole new generation of restored and rebuilt ex-World War 2 bombing aircraft. *Executive Sweet* became an immediate award winner and, in many ways, this single aircraft was responsible for beginning the trend of restoring warbirds as accurate representations of the way they flew in combat. Over the years, *Sweet* attended many airshows in North America and was always a hit with the crowds. During this period this aircraft—with its generous hatches and glazed areas—became a faithful mount for several aerial photographers. The Mitchell is now under the proud ownership of the American Aeronautical Foundation Museum. Note the open bomb-bay doors. [*Michael O'Leary*]

RIGHT A somewhat dubious Bruce Guberman looks on as this photographer pilots *Executive Sweet* through Southern California smog on the way to an assignment. Guberman, an FAA examiner in the B-25 and A-26 as well as being type rated in the B-17, was the photo plane pilot for many of the pictures featured in this book. [*Michael O'Leary*]

ABOVE Pushing the throttles for the R-2600s forward, Tom Crevasse breaks away from the camera plane and heads back to the Space Center Executive Airport during the May 1987 Valiant Air Command show. A few weeks later, Tom skilfully crash-landed another Mitchell in the Everglades swamp near the airfield. Both engines on the Mitchell failed, but by using his considerable piloting skills Tom got the bomber down with a minimum amount of damage to the airframe and no injuries. [*Michael O'Leary*]

LEFT Tight formation: Tom Crevasse pilots Harry Doan's camouflaged Mitchell in close formation with the Beech AT-11 camera plane. This aircraft had been an abandoned cargo freighter but was restored to flying condition by Doan and his crew in Florida. The Mitchell is a B-25J, registered N9621C (s/n 45-8811). With today's current values, a fully restored B-25 brings around $300,000 on the warbird market. [*Michael O'Leary*]

BELOW Beautifully polished Mitchell *Georgia Mae* (NL5262V) is owned and flown by Wiley Sanders of Troy, Alabama. Wiley must like Mitchells since he owns two of them – this glass-nose example and a solid nose 'strafer' variant. Most pilots who fly the Mitchell describe the aircraft as being like a giant Piper Cub – except when an engine quits! [*Michael O'Leary*]

ABOVE Turretless B-25J N9643C (s/n 44-86758, c/n 108-57512) is flown by the Confederate Air Force and done up as a United States Marine Corps PBJ-1J Mitchell operated by VMB-612. Surplus Mitchells have been used for a wide variety of tasks – as fire bombers, fish freighters and drug runners.

Fortunately between 20 and 30 Mitchells are still capable of taking to the air and this number is slowly growing thanks to such outfits like Aero Trade at Chino Airport, who not only restore Mitchells back to flying shape but also maintain a huge parts inventory from all around the world. [*Michael O'Leary*]

ABOVE Gear coming up, Canadian Warplane Heritage's B-25J climbs away immaculate in the colors of No 98 Sqn Royal Canadian Air Force. After many years as an executive transport, test bed and avionics demonstration aircraft with the Bendix Corporation the CWH flagship was bought for $12,000 and is one of two Mitchells in the CWH fleet. [*Mike Jerram*]

LEFT Based at Mt Hope, Hamilton, Ontario, the Canadian Warplane Heritage maintain one of the finest collections of flying World War 2 aircraft and the B-25J C-GCWM is a true star of their fleet. Since being acquired by the CWH in the mid-1970s the Mitchell has been a flying restoration, gradually being restored to World War 2 configuration as a Mitchell III while being kept operational. When this October 1976 photo was taken the CWH still had not added the upper gun turret or the full compliment of nose weapons. Today, the beautifully finished bomber reflects the pride and care that the CWE lavishes on its fleet of vintage warriors. The J was the most numerous variant of the Mitchell; more than 4000 were built, and it was capable of carrying more than 3000 lb of bombs. [*Michael O'Leary*]

B-25 MITCHELL

The movie version of Joseph Heller's famous novel *Catch-22* was responsible for saving the majority of America's Mitchells – most of which, by the time the movie began to be made in 1968, had lapsed either into dereliction or, at best, semi-flyable condition. For the film 18 Mitchells were made flyable, fitted with fake turrets and given fanciful military camouflage schemes. The bombers were then flown en masse to the filming location in Mexico. After their return to the Tallmantz facility at Orange County, California, where the rebuilding work has taken place, the bombers were sold off by the studio at prices ranging

from a mere $3000 to $6000! B-25J *Denver Dumper* is seen waiting for a buyer at Orange County in June 1969. Note the piles of Mitchell parts and spares surrounding the aircraft. Heller, by the way, was himself a Mitchell crewman in World War 2. [*Michael O'Leary*]

B-25 MITCHELL

BELOW, INSET Close-up of the well-done nose art on Ted Itano's *Pacific Princess*. This view also shows the swivel-mounted .50 caliber Browning machine-gun and Norden bombsight. TB-25Ns were Mitchells overhauled in the 1950s by Hayes Aircraft for the USAF to use as trainers. [*Michael O'Leary*]

BELOW It's not as easy as it looks! Formation flying in the Mitchell is a very demanding task, requiring constant attention to the heavy controls and throttles for the Wright R-2600s. On a hot and bumpy day over

California's central valley, Carl Scholl leads a pack of Mitchells in very tight formation. Painted as a US Navy PBJ-1J, the lead TB-25N (N9856C, 43-28204) is owned by Ted Itano. [*Michael O'Leary*]

BELOW This beguiling B-25 lady certainly did her very special *Show Me* routine over many enemy targets, bagging three Japanese fighter pilots (whose concentration must have been momentarily diverted) into the bargain. [*Norman Pealing*]

B-25 MITCHELL

BELOW With its left engine 'caged' *In the Mood* heads back to its home field at Chino, California. N9117Z is a B-25J, s/n 44-29199, s/n 108-32474 – that is basically a 'plain Jane' without turret, guns or much in the way of decoration except for some nose art. The plane is also shown on the previous page. [*Michael O'Leary*]

RIGHT One of the finest Mitchells belonged to Dr John Marshall. N10564 (B-25J, s/n 44-29887, c/n 108-33162) is a *Catch-22* veteran and ex-Forest Service fire bomber, but thousands of man hours and lots of money created a fine replica of a combat-operational wartime bomber. [*Michael O'Leary*]

ABOVE Painted in attractive post-World War 2 Royal Canadian Air Force No 418 'City of Edmonton' Squadron, this B-25J, 45-8884, was photographed over Madera, California during August 1982. Owned by Jerry Janes and registered in Canada as C-GCWJ, the aircraft had previously been N3156G. The RCAF acquired 75 B-25Js from surplus USAF stocks in 1951 in order to supplement 70 Mitchell IIIs that had been obtained previously from the RAF. These B-25Js saw service until 1963 – about the time the USAF was retiring its last Mitchells, used as personnel transports and trainers. [*Michael O'Leary*]

ABOVE During World War 2 the USAAF's famous 'Air Apaches' caused constant havoc among Japanese shipping and airfields in the Pacific. Skimming in at tree-top or wave-top height, the Mitchells of the Air Apaches hit the enemy with withering concentrations of .50 caliber machine-gun fire and loads of fragmentation bombs. This Mitchell, owned by Tom Thompson, is restored in dramatic Air Apache colors. Note the operational top turret tracking the camera plane! [*Michael O'Leary*]

LEFT The distinctive 'gull-wings' of the Mitchell came about as as result of the need to improve directional stability by eliminating dihedral outboard of the engine nacelles. Whatever the aerodynamic reasoning, the modification gave the B-25 a much more purposeful appearance. [*Norman Pealing*]

BELOW Americans and Texans both: Stars and Stripes and Lone Star flags flutter over a beautifully restored North American B-25 Mitchell of the Confederate Air Force, also featured in the other pictures here. One of the great bombers of World War 2, the B-25 was named after Brigadier General Billy Mitchell, a pioneer advocate of military air power. Five B-25s and one Navy PBJ-1C now fly with the CAF. [*Norman Pealing*]

LEFT *Yellow Rose* must have been a Texas lady often taken into battle on this B-25. Here's hoping her cowgirl boots and stetson helped the *good ole gal* maintain some sort of decorum in times of stress . . . [*Norman Pealing*]

As the sun rises on another day that promises to do good things for Budweiser stockholders, the heady delights of looking and listening on the Harlingen ramp become apparent. A Mitchell coughs into life, producing a rich-mixture smoke haze – can't you just smell it! Only big radials can sound like this as *Devil Dog* quivers expectantly on her fat tyres before the pilot eases off the brakes, adds a touch of throttle and rolls the bomber towards the taxiway. [*Norman Pealing*].

Medium marvel 2: INVADER

With one of the longest service lives of any American combat aircraft, the A-26 was designed, developed and produced within World War 2, flying for the first time as the XA-26 on 10 July 1942 and making its debut with service units in December 1943. Nearly 2500 Invaders had been built when the whole program was terminated after VJ-Day, the final delivery occuring on 2 January 1946.

The Invader's service record in World War 2 was exemplary, dropping 18,000 tons of bombs in 11,000 sorties for the loss of only 67 aircraft. It went on to serve with distinction in Korea and, remanufactured by On-Mark as the B-26K (A-26A), performed night interdiction sorties along the Ho Chi Minh trail in Vietnam. The A-26A could carry a warload of 11,000 lb at 350 mph and loiter for two hours in the target area.

When the Martin B-26 Marauder was withdrawn from service in 1948, the A-26 was redesignated as the B-26. During the 1970s, Invaders could be purchased for small amounts of money. Their high top speed and long range made them ideal candidates for drug runners. Numerous aircraft were either wrecked, impounded or abandoned while engaged in this activity.

RIGHT This magnificent Douglas A-26 Invader came all the way from Canada to attend the Confederate Air Force's 20th anniversary at Harlingen in 1987. Of such stuff are true believers made! [*Norman Pealing*]

ABOVE Nose job: close-up of the Invader's distinctive front end. [*Norman Pealing*]

RIGHT B-26C 41-39401 *Whistler's Mother* was purchased by film-maker Dick Moore when the Invader was surplused in 1958. The plane was flown to Van Nuys Airport in California, where it became a local but decidedly non-flying landmark. During the 1960s the Invader was put back into the air briefly for its movie role. The interesting point about this aircraft was the fact that it had never been demilitarized: all the original military equipment, including the turrets, was still fully operational. During the late 1970s the bomber attracted the attention of Ed Schepf and the American Aeronautical Foundation Museum. The aircraft was

transferred to the AAF and an intensive restoration project was undertaken – a task comprising five years and 5000 man-hours. Head of the detailed restoration was Nelson Knuedeler, who did an outstanding job insuring that every function of the bomber was brought back to military standard.

N39401 flew 30 combat missions in Europe with the 643rd Bomb Squadron, 409th Bomb Group, before being returned to the States following VE Day. Its peacetime rest was to be brief, however, and the Invader (redesignated B-26 in 1947) went to war in Korea – flying a further 100 combat missions. Returned once again to the States, the B-26C was eventually transferred to the 180th Tactical Reconnaissance Squadron of the Missouri Air National Guard at St Joseph. [*Michael O'Leary*]

INSET (LEFT) The heart of the Invader's incredible reserve of power: two mighty Pratt & Whitney R-2800 radials capable of giving a top speed of 355 mph at 15,000 feet. Very reliable engines, the P & Ws were one of the main reasons that the Invader came out of the European air war losing only 67 of its type in combat. [*Michael O'Leary*]

INSET (RIGHT) The Douglas Invader combined fighter-like performance with excellent visibility – as seen here as Bruce Guberman takes *Whistler's Mother* down low over the Pacific at 275 mph. Though quite a large aircraft the Invader's cockpit is cramped, especially the jump seat from which this photograph was taken. [*Michael O'Leary*]

A-26 INVADER

BELOW N39401 gracefully breaks away from the camera aircraft. Aside from being involved in three major wars (World War 2, Korea, Vietnam), Invaders have seen action in many of the world's hot spots

including mercenary action in Africa, numerous Latin American 'problems' and in the infamous Bay of Pigs invasion. Today a stock Invader is an extremely rare warbird since the majority of surviving airframes tend to be various 'flavors' of the many different executive modifications done in the 1950s and early 1960s. *Whistler's Mother* will probably remain as the prime example of an operationally restored Invader. In late 1985, ownership of the Invader passed to Kermit Weeks and the aircraft is now on display in his newly opened Weeks Air Museum, Tamiami Airport, Miami, Florida. [*Michael O'Leary*]

RIGHT With the massive bomb doors open, the Invader makes for an impressive sight. The doors could be opened right up to red line speed. Note the barely visible spoilers in front of the doors that automatically lower as they open. The spoilers break up the airflow in front of the doors, making sure that the bombs can fall freely from the bay. On later variants of the A-26C (from the C-45-DT block on), the Invader had three .50 calibers mounted internally in each outer wing panel. [*Michael O'Leary*]

ABOVE, INSET This view shows the bottom turret on the A-26C to advantage. With its high top speed (airframe was redlined at 425 mph, but this was quite often exceeded in the heat of combat) and heavy armament, the Invader often met and defeated enemy fighters. [*Michael O'Leary*]

With its huge tricycle gear lowered, *Whistler's Mother* practices slow flight shortly after restoration with Bruce Guberman at the controls. At this point (August 1983) the four underwing gun packs (each housing two .50 caliber machine-guns) had not been installed. [*Michael O'Leary*]

ABOVE The twin turrets on the Invader were remotely controled by a gunner sitting under a glass enclosure to the rear of the fuselage. The gunner operated the electrically-driven turrets by a pair of handgrips attached to a periscope sight (the top of which can be seen projecting from the gunner's position). The A-26C packed up to 16 machine-guns during World War 2 and Korea. [*Michael O'Leary*]

TOP Gun-nose A-26B Invaders await their airshow flights at the annual Breckenridge, Texas warbird gathering in 1983. N240P, in the foreground, is finished in the overall black night intruder markings while N26RP is finished in glossy Olive Drab. Both aircraft had been converted into executive transports during the late 1950s or early 1960s, as can be seen by the airstair entrance doors located in the area formerly occupied by the bomb-bay. Lots of money and volunteer labor were poured into N250P to bring it back to flying condition. [*Michael O'Leary*]

ABOVE Polished like a mirror, VB-26C 0-34610 is seen on the ramp at Edwards AFB during May 1972. Operated by the National Guard Bureau out of Washington DC, this Invader (along with a couple of others) was employed as a VIP transport for ranking Air National Guard officers. Declining availability of parts and increasing difficulty in obtaining avgas forced the retirement of these classics by the mid-1970s. [*Michael O'Leary*]

ABOVE Ready to race. During the late 1950s and 1960s On-Mark Engineering at Van Nuys, California established itself as the leading modifier of Douglas Invaders into executive transports. On-Mark completely redid the basic airframe, adding new ring spars which permitted higher operational speeds as well as allowing passengers to stand more or less upright in the cabin. CB-17 engines added super high horsepower while the deluxe versions had cabin pressurization, providing the ultimate in pre-Learjet executive transportation. Once the jets arrived the On-Marks were quickly dumped onto the open market, usually at very low prices. A young Pan American pilot by the name of Lloyd Hamilton purchased On-Mark N500M and raced it at the 1971 Mojave air races, where the powerful twin did well. Race 16 is seen ready for the next round at Mojave with its executive nose and tip tanks. [*Michael O'Leary*]

LEFT Considered to be the ultimate Invader, a row of On-Mark B-26K Counter Invaders is seen awaiting scrapping at Davis-Monthan AFB during September 1971. Stock Invaders held in storage by the USAF were completely redone at Van Nuys, California, by On-Mark during the early 1960s for operations in Southeast Asia. New, powerful R-2800 CB-17 engines were installed along with new spars, strengthened wings and new avionics. The planes were given new serials (this example being 64-17671) and then flown to Southeast Asia to be employed on a variety of heavy-duty missions against the Communists – many of the missions being clandestine. After their military use, the government did not want the planes to be sold surplus to civilians and most of

the battle-hardened veterans were cut up for scrap metal. However, a couple did survive and one Counter Invader has been restored to operational condition in Montana. This particular machine went to the Forest Service, then to the Florence Air Museum, South Carolina. [*Michael O'Leary*]

BELOW LEFT Well-armed guards puzzle on why some crazy gringos want to fly this marginally airworthy Invader out of Tegucigalpa, Honduras, during December 1982. This aircraft was the only Invader used by the *Fuerza Aera Hondurena* and was eventually put up for sale following the acquisition of more modern equipment. This Invader, s/n 44-35918 c/n 29197, was obtained from Costa Rica in 1969 where it had been carried on the civil register as TI-1040P. Given the Honduran civil register of HR-276, the attack bomber was transfered in 1971 to the FAH as 276, later FAH-510. [*Michael O'Leary*]

ABOVE After sitting idle for several years, the R-2800s on FAH-510 burst into life on the first try. After several test flights at Tegucigalpa, Mike and Dick Wright and Dave Zeuschel flew the Invader to Texas where it was put on permanent static display at a USAF base. The vintage bomber performed quite well on its ferry flight which was just as well – some of the territory flown over was inhabited by rather primitive natives who had the occasional hankering for a nice chunk of human. [*Michael O'Leary*]

Top Two classics bask in the Arizona sun. Invader B-26C N9996Z receives some work to the tail cone area while the mechanic's '55 Cadillac stands guard. Photographed at Deer Valley Airport in April 1968, the Invader was being operated as a fire bomber with the Forest Service area code 11C painted on the vertical tail. The B-26 has enjoyed a long life as a fire bomber, the power from the Pratt & Whitney R-2800s being particularly appreciated by pilots operating with heavy chemical loads in very hot environments. [*Michael O'Leary*]

Above Updated to B-26K configuration, *Forca Aerea Brasileira* B-26C (note that the old designation was retained) 5174 awaits an engine run at Hamilton. Camouflaged glossy green and gray, the Invader has two underwing weapon pylons as well as rocket rails. Wings are fitted for six .50 caliber Browning machine-guns. Pictured in May 1969, the rebuilt Invaders went on to serve Brazil for ten years before being retired. [*Michael O'Leary*]

184

Top Unless a surplus bomber had a very specific use or caring owner, most surviving examples were rapidly heading downhill during the early 1960s. The first executive jets were coming on the market, so many transport conversions were being dumped in favor of the more practical Learjets. This Douglas A-26B is seen in a sad state at Fresno, California. Basically in stock condition, the Invader still retains its overall black night intruder camouflage, the white being added by civilian owners. [*Michael O'Leary*]

Below In the late 1960s Brazil decided to have its fleet of Invaders upgraded to basic B-26K configuration. By this time On-Mark had gone on to other projects so a license for the conversion work was issued to Hamilton Aircraft in Tucson, Arizona. The Brazilian Invaders were ferried north and modifications were undertaken on a production line basis. Stripped of its engines, B-26C 5159 is seen here awaiting conversion. Note the underwing gun pods and rocket rails. [*Michael O'Leary*]

INSETS Step-by-step guide to approach work. Low over the Texas scrub, *Puss & Boots* settles in for its final approach, nose held high and flaps full down; lined up on the centreline at Breckenridge, approaches the threshold; throttles come back to 15 inches manifold pressure and speed to 105 mph and prepares to join its shadow. [*Michael O'Leary*]

BELOW Discovered in the early 1980s in a derelict condition in Brazil, this Invader was bought by the Tired Iron Racing Team in Casper, Wyoming. After purchase, only one problem remained . . . getting the plane back to the United States! So Mike and Dick Wright traveled to Brazil, where they spent months bringing the old bird back to life. The brothers put the aircraft into flyable shape with the addition of two new R-2800s and lots of labor. This aircraft had served with the *Forca Aerea Brasileira* and was one of the Invaders modified by Hamilton Aircraft in Arizona. After passing from active service, the airframe was basically abandoned. Fortunately, the ferry flight back to the States went smoothly and the Invader's restoration was completely finished. The resulting fine product, named *Puss & Boots*, is seen in this photograph turning final at Breckenridge. Gear speed on the Invader is 160 mph and the final approach is flown at 130 mph with flaps full down. [*Michael O'Leary*]

Medium cool: HARPOON, MARAUDER & HAVOC

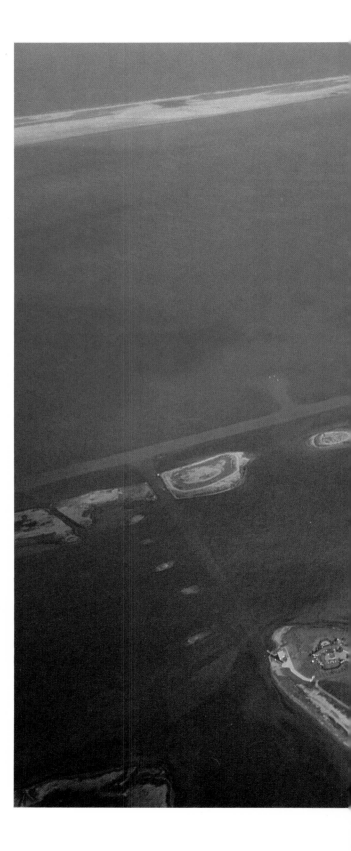

The big four-engined heavies of the American war machine have always grabbed the head-lines when it comes to the winning of World War II, but a supporting cast of extremely effective light-medium bombers from several noted US companies also swayed the conflict in the Allies' favor.

The purposeful Martin Marauder per-formed sterling low-level work over North Africa and Europe. A naval equivalent of the Marauder was the Lockheed Ventura/ Harpoon; developed from the Hudson, the Lockheed twin was extensively used in the Pacific against Japanese shipping, as well as against German ports and coastal convoys in northern Europe.

The first of the light twins to enter service, the Douglas A-20 Havoc was the workhorse for many Allied squadrons – including those of the Soviet Union, who purchased almost half the total Havoc production. A total of 7385 Douglas twins were constructed but, like the Marauder, only a handful are still with us.

RIGHT The Lockheed PV-2 Harpoon was a redesigned version of the undistinguished Ventura bomber built for the RAF in 1941–42. One of the 'limited editions' of World War 2, only 535 Harpoons were built, but as a long-range torpedo bomber for the US Navy the aircraft was used with considerable success in the South Pacific theater. [*Norman Pealing*]

BELOW Richard and Maggie Mitchell spotted *Fat Cat* as a derelict hulk just off the runway at Marianna Airport, Florida in August 1984. After a superb restoration effort, *Fat Cat* took to the skies again in March 1985. [*Norman Pealing*]

RIGHT Unlike the Ventura bomber, the PV-2 Harpoon was heavily armed. Up to ten forward-firing .50 caliber guns could be carried in the nose and 4000 lb of bombs or torpedoes internally, plus a further 2000 lb of stores under the wings. With this warload the Harpoon had a range of about 1800 nautical miles cruising at between 125–135 mph – and allowing for acceleration to around 300 mph for combat. [*Norman Pealing*]

BELOW The Harpoon was also used as a make-do long-range fighter in the Pacific and claimed the destruction of a number of Zero-Sens. Other interesting duties included fighter-bomber escort for Army B-24s in the Aleutians as well as C-47s on paradrop missions in New Guinea. [*Norman Pealing*]

PV-2 HARPOON

BELOW *Fat Cat* is a restored Lockheed PV-2 Harpoon with an interesting history. N7428C, after being surplused from US Navy service, went through several different owners before being sent to Dee Howard Aircraft in Texas for extensive modification. The new owners wanted the Harpoon heavily modified for use as a freighter to bring rare plants from Latin America to the States. Accordingly contracted in 1965 to Dee Howard, a leading aircraft modifier, the Harpoon's interior was gutted and a large cargo door was cut into the right side of the fuselage. The most interesting aspect of the modification was the fact that the fuselage was stretched by four feet. [*Michael O'Leary*]

Left, inset N7428C's career as a freighter was short-lived. The modification was extremely expensive, around $200,000, and the market for rare plants apparently not as extensive as the new owners thought. After passing through more owners, the Harpoon began a new career as a drug runner. Eventually impounded, the PV-2 was purchased by Richard Mitchell of New Iberia, Louisiana. A group of volunteers active in the Cajun Wing of the Confederate Air Force began getting their big bird

ready to fly again and this was no small task since the Harpoon had been more than mistreated. The airplane was eventually put into ferry condition and has been the subject of a working restoration. As can be seen in our photos, *Fat Cat* has been nicely polished and some military details like machine-gun barrels and underwing rockets have been added. Flown by Richard Mitchell and Bob DeRosier, the bomber is now a popular airshow regular. [*Michael O'Leary*]

Above One of the most neglected of all warbird bombers is the Lockheed PV-2 Ventura, 500 of which were built at Burbank, California, starting in mid-1943. The Harpoon was a heavy-hitter, built for the US Navy's attack bomber role. Carrying up to ten .50 caliber machine-guns, eight 5-inch rockets and a heavy bomb-load, Harpoons attacked and destroyed Japanese targets all across the warfront – quite often engaging and defeating the vaunted Mitsubishi Zero. Used by the Naval Reserves after the war, most Harpoons quietly disappeared to the scrappers, but

many were converted to high-speed executive transports while others were employed as fire bombers and bug sprayers. Only catching on as a collectable warbird in the early 1980s, around a half dozen Harpoons are currently under restoration to full military configuration. This example (N7255C, BuNo 37257) was quietly rotting away with several other Harpoons at Douglas, Arizona, in 1970 – unfortunately indicative, at the time, of the interest in a warplane that had seen considerable action in the defense of its country. [*Michael O'Leary*]

LEFT This Martin Marauder has been a hard-luck warbird but it managed to survive a number of misfortunes. Surplus immediately after the conclusion of the war, N5546N was turned into a cross-country racer and flown as the *Valley Turtle*. Not particularly successful, the B-26 was then sold for conversion to executive status and the modifications were extensive, including a new nose and tail section, deluxe interior, airstair door and new props and cowlings. Used by an oil company, the plane was later sold in Mexico as XB-LOX. Returned to the States in the late 1960s, the aircraft was damaged in a landing accident and acquired by the CAF. After being moved to Harlingen, the Marauder was very heavily damaged when the gear retracted while rolling (note the bent wing and other damage). After being stored, the CAF raised funds to try to restore the Marauder back to wartime configuration. It took years but the project was completed–only to have the nose gear collapse on touchdown during a 1986 show. The aircraft has again been repaired and is the only flying example of Martin's bomber. [*Michael O'Leary*]

ABOVE Chino Airport in Southern California is the home of many exotic warbirds. David Tallichet's Yesterday's Air Force holds upwards of 150 aircraft and airframes awaiting restoration. Perhaps one of the rarest aircraft in Tallichet's monumental collection is this Martin B-26 Marauder. One of three recovered by Tallichet from the wilds of Canada in the early 1970s, the Marauders had gone down after becoming lost on a ferry flight and running low on fuel. One was badly damaged in the accident but the other two remained relatively undamaged, the cold Canadian climate keeping the metal and paintwork in a remarkable state of preservation. One example was selected for restoration after having been transfered back to Chino and the short-wing Marauder is seen as restoration began in 1977. Today, the aircraft is the world's only airworthy Marauder.
[*Michael O'Leary*]

RIGHT Upward-opening cockpit roof halves are a characteristic of both the Douglas A-26 Invader and the Martin B-26 Marauder. This is *Carolyn*, owned by the Confederate Air Force and the only Marauder currently maintained in airworthy condition from a wartime production run of 5266 aircraft. [*Norman Pealing*]

ABOVE From the start of its service career the B-26 was regarded as a hot ship, thanks to its high wing loading and abundant power. Training accidents were high and 'The Widow Maker' came to be feared by inexperienced pilots. In the right hands the Marauder presented few problems, but Martin bowed to the inevitable and extended the wingspan and vertical tail.

In May 1943 the modified B-26B was introduced to the 9th Air Force in Europe and the Marauder rapidly dispelled any remaining doubts about its safety record and operational suitability. By VE-Day B-26 units had suffered combat losses of less than one per cent – the lowest loss rate of any US Army bomber in Europe. [*Norman Pealing*]

ABOVE Owned and operated on behalf of the Confederate Air Force, *Carolyn* returns from its display after playing its part in celebrating the 30th anniversary of the 'Ghost Squadron' at Harlingen in 1987. This sexy lady was acquired in 1967 as a distinctly tired corporate transport, and in 1975 the CAF formed the 'B-26 Squadron' under Colonel Jerry Harville. A total of nine years and $250,000 later, *Carolyn* was rightfully restored to Army Air Corps configuration. [*Norman Pealing*]

ABOVE Sting in the tail: in common with many other American bombers, the B-26 was armed with a pair of .50 caliber Browning machine-guns in a powered tail turret. [*Norman Pealing*]

ABOVE One of the hardest-working attack bombers of
World War 2 was the Douglas A-20 Havoc series,
fighting with American forces all the way through the
conflict. However, no example of the aircraft remains
flyable (out of 7098 examples built by Douglas and
380 by Boeing) while only a handful are on display in
museums around the world. Found as a derelict
airframe by William Farrah of El Paso, Texas, this
A-20G required several years of intensive labor to get
back into flying condition. The A-20G variant of the
attack bomber is powered by two Wright R-2600
radials and is capable of a top speed of 315 mph. First

flown on 26 October 1938 (Douglas B), the Havoc was
a versatile, hard-hitting aircraft. The French
government ordered 100 DB-7s in February 1939 and
the first of these went into action against the invading
German forces on 31 May 1940. After the fall of
France, the few DB-7s which escaped to Britain were
supplanted by fresh deliveries to the RAF and these
were used in both bomber (as the Boston) and night-
fighter roles. RAF pilots who flew the French A-20s
diverted to Britain had to grapple with metric
instrumentation and 'opposite' throttles (i.e. forward
to throttle back and vice versa). [*Michael O'Leary*]

RIGHT Havoc with a hammock: not a shot from a CAF holiday brochure but a veteran 'Colonel' reflecting(?) on the Havoc's glorious past. It was found forgotten and ailing in Boise, Idaho, with engines not run for 12 years; all of its fabric control surfaces were completely rotten. After many months of restoration it was very cautiously flown to Rebel Field and faithfully restored in the colors of south-western Pacific units and the 312th Bomb Group based in New Guinea, where swashbuckling low-level anti-shipping operations became their calling card with the Japanese. [*Norman Pealing*]

BELOW Bomber restorer's paradise! This 1980 view shows just a portion of David Tallichet's vast warbird holdings at Chino, California. Tallichet, a collector who has traveled the world in search of rare and exotic World War 2 aircraft, has restored a number of planes to flying condition but, unfortunately, none of the aircraft in this view. The disassembled Douglas A-20 Havoc in the foreground was recovered from Nicaragua shortly before the current *Sandinista* regime clamped down on most activities (including, one would surmize, the collecting of old warbirds). Records fail to show any evidence of Havocs operating with the *Fuerza Aerea de Nicaragua* and it would be interesting if the vintage attack bomber could talk. Other aircraft in the background include a vandalized Mitchell from *Catch-22*, a Bristol Bolingbroke, an ultra-rare Curtiss SB2C Helldiver and a TBM-3 Avenger. [*Michael O'Leary*]

LEFT This piratical design adorning the Confederate Air Force Havoc incorporates the barrels of the four 20 mm cannon carried in the nose of the Douglas A-20G. Tragically this aircraft, the only airworthy example of its type in existence, was destroyed in 1989 after its pilot suffered a fatal heart attack at the controls. [*Norman Pealing*]

ABOVE The Havoc certainly lived up to its name during operations against Field-Marshal Erwin Rommel's *Afrika Corps* in the North African campaign of 1942–43, when aircraft of the US Army 9th Air Force bombed and strafed the troops and impedimenta of 'The Desert Fox' across Morocco, Tunisia and what we know today as Libya. Later, the 9th Air Force used their Havocs with similar success to help nullify the Axis defenses before the Normandy invasion. [*Norman Pealing*]

BELOW All the later marks of Havoc, including the
A-20G seen here, were powered by Wright GR-2600
Cyclone 14-cylinder radials of up to 1700 hp.
Production had reached over 7000 of all marks when it
was terminated in September 1944. A total of 3125
A-20s were supplied free to the Soviet Union during
World War 2. [*Norman Pealing*]

Heavyweight champion 1: FLYING FORTRESS

The legendary B-17 was perhaps the greatest symbol of America's military effort against the Axis powers in Europe, when thousands darkened the skies over the occupied countries as the Eighth Air Force's daylight bombing campaign went into top gear. Although outnumbered by its four-engined partner, the B-24 Liberator, and possessing a bomb-load that was less than awesome, the legendary deeds of the Eighth's aircraft and their crews secured a permanent place for the bomber in aviation lore.

The development of Model 299, as it was called by Boeing, into the definitive 'warwagon' took almost a decade, the prototype flying for the first time on 28 July 1935. The ultimate Fortress, the B-17G, did not find its way to front-line squadrons until September 1943, no less than five sub-types preceding it. A total of 12,751 Flying Fortresses eventually took to the skies – of which more than a third were lost in combat – but, unlike its fellow heavy-hauler of the European theater, the B-24, the Fort is still found in some numbers.

LEFT Daybreak at Harlingen highlights the Confederate Air Force's very own Flying Fortress. The CAF's Fort is a B-17G, the definitive production version; Boeing turned out 4035 at its Seattle plant, with Vega and Douglas adding 2250 and 2395. [*Norman Pealing*]

B-17 FLYING FORTRESS

BELOW During 1986 Aviation Specialities, a large fire bombing and aerial applicator company, decided to auction off its collection of vintage workhorses. Included in this collection were four operational B-17 Flying Fortresses. Fortunately, all these aircraft were purchased by caring collectors who are now bringing the machines back to stock World War 2 condition. One of the most ambitious groups is the National Warplane Museum in Genesco, New York, who obtained N9563Z (s/n 44-83563, c/n 32204) and flew it back to their home airfield where the aircraft is being kept in operational condition while restoration work continues. [*Michael O'Leary*]

BELOW, INSET N9563Z cruises over the Canadian countryside. As can be seen, the nose glass was in extremely poor condition when the plane was purchased and this will eventually be replaced. The National Warplane Museum hopes to obtain a complete selection of gun turrets and will eventually paint their big bird in an accurate World War 2 scheme. [*Michael O'Leary*]

BELOW Its fading and chipped civilian paintwork stripped off and its aluminum skin polished, N9563Z made one of its first airshow outings at the June 1986 Canadian Warplane Heritage event. By this time, an early model top turret had been installed and much mechanical work had been carried out. During its career a cargo door had been installed in the right side of the fuselage and this will be removed. N9563Z is a veteran of the 1961 film *The War Lover* as well as the later *Tora! Tora! Tora!* [*Michael O'Leary*]

B-17 FLYING FORTRESS

BELOW An unusual view of a Flying Fortress in formation. In October 1977 Aero Union, a fire bomber outfit based in Chico, California, brought one of their Forts, N9323Z, to the Confederate Air Force annual show at Harlingen, Texas. In hopes of selling the veteran bomber, Aero Union applied some stars and bars along with the name *Class of '44* (the large red areas are visibility aids during retardant drops). Asking price was in the vicinity of $55,000 for the fire bomber, which was in anything but stock condition, but Aero Union's Forts were always maintained in superior condition and the price was more than a bargain. N9323Z was sold and, of course, went on to become the sensational *Sentimental Journey*. This view shows the Fort with every nook and cranny filled with passengers. Note the top hatch from the radio compartment has been removed and the passengers are standing up in relative comfort, the hump of the top fuselage apparently blocking the airflow. [*Michael O'Leary*]

BELOW LEFT *Sentimental Journey* prepares for a low pass at the 1982 Madera Gathering of Warbirds. If there is one single aircraft that captures public imagination when bombers of World War 2 are discussed, then that aircraft is the Boeing B-17 Flying Fortress. [*Michael O'Leary*]

BELOW When obtained by the Arizona Wing of the CAF the B-17G was without turrets, but diligent work by the CAF has seen all three restored to the airframe. Once common in most surplus scrapyards, turrets have now become sought-after collector's items, with prices ranging as high as $10,000 or more. [*Michael O'Leary*]

ABOVE The fine condition of *Sentimental Journey*, probably the most authentically restored Fortress still flying, is reflected in this view showing its aluminum skin glowing from repeated polishings by the dedicated volunteer workers. [*Michael O'Leary*]

ABOVE LEFT Big friends, little friends: a pair of P-51D Mustangs provide top cover for *Sentimental Journey* just like it was back in '44. [*Mike Jerram*]

LEFT AND BELOW You can tell why the little guy was always the ball turret gunner in any Flying Fortress crew. Hunched embryo-style on a bicycle-saddle seat and sighting his twin .50s through his raised legs, his was the bomber's most important and seemingly most vulnerable defensive position, yet statistically it proved the safest crew station in combat. Other crew members had to wind the turret up into the fuselage to enable the ball gunner to leave his cramped quarters. [*Mike Jerram*]

B-17 FLYING FORTRESS

Gear down and in a steep bank while turning final for landing, *Sentimental Journey* makes an imposing sight in this May 1985 photograph. Based at Falcon Field, Mesa, Arizona, the aircraft regularly visits airshows across America. A modest fee is charged for tours of the interior of the famous bomber and this money goes a long way towards keeping the immaculate veteran flying. *Journey's* nose art depicts America's most famous World War 2 pin-up, Betty Grable. The B-17G (s/n 44-83513, c/n 32155, N9323Z) is owned and operated by the Arizona Wing of the Confederate Air Force. [*Michael O'Leary*]

ABOVE AND RIGHT One of the most recent Fortress restorations is Bob Collings's magnificent Boeing B-17G N93012, painted up as *Nine-O-Nine*, a World War 2 Eighth Air Force Fort based in Britain. Last operated in USAF service by the Military Air Transport Service, the Fort was purchased by Aviation Specialties of Mesa, Arizona, and put in flying shape for a ferry flight back to Falcon Field. Dubbed *Yucca Lady*, the Fort was rebuilt over a period of ten years by Gene Packard and his crew. 'We rebuilt the aircraft exactly to fit our operations,' recalled Packard. 'All the military equipment was junked.' Once finally finished, the bomber was used on forest fires and for spraying. Sold off in the 1986 auction, the Fort went to Bob Collings, who immediately had the plane flown to Tom Reilly's 'Bombertown' near Orlando, Florida. Reilly and his large crew went to work to bring N93012 back to military condition and the result can be seen here. A top turret was still being completed and had not been installed by the time this March 1987 photograph was taken. [*Michael O'Leary*]

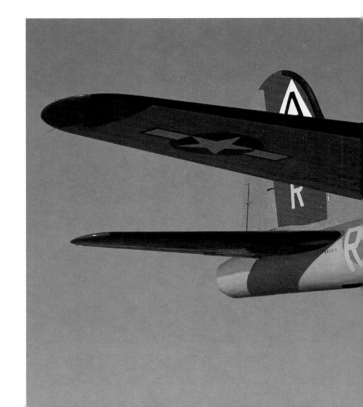

212

LEFT David Tallichet flying his B-17G during August 1985. Tallichet is the only current B-17 owner that actually flew Forts during combat, operating out of Thorpe Abbots in England with the 100th Bomb Group of the Eighth Air Force. N3703G had been operated by TBM Inc as a fire bomber before being obtained by Tallichet's Yesterday's Air Force. The Flying Fortress has been painted up as an F model and is minus the G's characteristic chin turret. N703G is s/n 44-83546, c/n 32187. [*Michael O'Leary*]

RIGHT War – what war? A sight inconceivable 45 or so years ago! The CAF's B-17 in company with one of the 'Ghost Squadron's' *Kate* torpedo bombers and three *Zeke* replicas. These last four, with others, faithfully represent the 'baddies' at the annual re-enactment of the attack on Pearl Harbor. As one hardened veteran was overheard to say at the spectacle in Texas: 'Jeez, it's more scary here than those two hours in the Pacific all those years ago!' As for the B-17 (which plays the part of a crippled 'Fort' in the demonstration) a comment by General Curtis Le May about the Eighth Air Force daylight bombing missions over Germany: 'None were ever aborted or turned back by enemy fighters or flak.' [*Norman Pealing*]

ABOVE Forts have returned to the United States from some exotic foreign locales. This B-17G last saw service with the Brazilian Air Force, where it was used for search and rescue (SAR) missions. Maintained over the years in good condition, the Fort was finally retired and sat for several years before being obtained by David Tallichet's Yesterday's Air Force. It's photographed at Chino, California, during October 1977, s/n 44-83663 (FAB 5400) where, with crew door open, the aircraft sits ready for a flight. [*Michael O'Leary*]

BELOW AND BELOW RIGHT That's how they do things in Texas, with more – well, *pizzazz*. The Confederate Air Force's B-17F *Texas Raiders* taxies out with Old Glory and the flag of the Lone Star state sharing honours from the flight deck windows . . .

ABOVE The B-17G owned by the Experimental Aircraft Establishment's Air Museum scarcely lives up to its name here, lacking the turrets and 13 .50 caliber machine-guns that provided the Fortress's formidable defenses. [*Mike Jerram*]

LEFT Four Wrights a-rumbling, crew watching the wingtips, the EAA's B-17G *Aluminum Overcast* eases out of a tight parking spot for a fly-by spot at the annual Fly-In and Convention at Oshkosh, Wisconsin. [*Mike Jerram*]

. . . and returns for a battle-damaged one wheel up, one wheel down touch and go which deservedly gets the good ole boys a round of applause, even from Damn Yankees. Just wait till they try a belly landing. [*Mike Jerram*]

ABOVE With bits of grass and associated airshow debris being kicked up by the props on its outboard Wright R-1820 radials (1200 hp each), the

Experimental Aircraft Association's N5017N swings around into its parking spot following an early morning sortie. [*Michael O'Leary*]

LEFT A spectator's view of the Experimental Aircraft Establishment's B-17G N5017N (s/n 44-85740). [*Mike Jerram*]

BELOW Climbing out from Titusville, Florida in a nose-up attitude, N5017N displays the Fortress's fine lines. Donated to the EAA by a group of warbird pilots who had purchased the ageing veteran from Dothan Aviation (where it served with N3701G as a fire bomber), the four-engine heavy is maintained in fine flying condition by the EAA. The prototype of the B-17 made its first flight on 28 July 1935. [*Michael O'Leary*]

BOTTOM The graceful lines of the B-17G will eventually be enhanced by the addition of turrets and other military equipment as the EAA finds time and funds to furnish their four-engine bomber with original World War 2 equipment. Most civil Forts were previously operated as the fire bombers until a Forest Service directive stated that the planes were too old for such hazardous operations. This released the bombers for sale to warbird enthusiasts but, unfortunately, many perfectly fine and flyable civilian B-17s went to the Air Force, where they now rot outside as gate guardians at the various bases that are part of the USAF Museum's heritage programme. Operators received 'newer' equipment (C-118s, C-123s, etc) for conversion to fire bombers via these trades. [*Michael O'Leary*]

B-17 FLYING FORTRESS

BELOW B-17G N3701G (s/n 44-8543) was operated for many years by Dothan Aviation in Alabama as a fire ant sprayer. These deadly ants infest the South and South-western US and regular dosings of poison from Dothan's two Fortresses helped decrease the threat to the public. As the two Forts became more uneconomical to operate, Dothan decided to sell both planes. Colonel 'Doc' Hospers purchased N3701G and, after much work, ferried the plane back to Fort Worth, Texas. Naming it *Chuckie* after his wife, Hosper has kept the Fort in flying condition while adding military bits and pieces to make it more authentic. Photographed at Breckenridge, Texas during May 1984, this view shows interesting details such as bomb-bay and covering patch over which the belly turret was installed. [*Michael O'Leary*]

LEFT Here's *Chuckie* three years later (May 1987), skirting around rainstorms surrounding Breckenridge. During the intervening three years, *Chuckie's* crew has accomplished much – adding a top turret, polishing the airframe, adding a new nose Perspex and installing much original military equipment in the interior. [*Michael O'Leary*]

ABOVE Colonel 'Doc' Hospers – he's a physician from Dallas – in his B-17 *Chuckie*, resplendent in polished aluminum. It's his own personal transport. Here he taxies for take-off after a visit to the Confederate Air Force at Harlingen, where the 'Ghost Squadron' celebrated its 30th anniversary in 1987. [*Norman Pealing*]

The fabulous *Sally B*, Britain's only operational Flying Fortress, is seen on an outing from its Duxford base, operated as an extension of the Imperial War Museum and located near the university town of Cambridge. Accompanied by two warbird fighters from Stephen Grey's ever growing fleet (Stephen is flying his P-51D while Steve Hinton is at the controls of the Thunderbolt), *Sally B* is being flown by Keith Sissons. B-17G-105-VE was obtained from USAAF stocks by *Institut Geographique National*, Creil, France, where, along with twelve other Forts, the vintage B-17 was put to use in a mapping capacity. Registered F-BGSR, the Fort was obtained by the late Ted White on 20 January 1975, registered N117TE and operated on the airshow circuit. Later registered G-BEDF, the B-17G is now maintained by a preservation trust and is enthusiastically flown as the last British example of a type of aircraft that once literally darkened the skies over the island nation

LEFT The beautiful B-17 continues to grace the skies over Britain decades later in the shape of *Sally B*. [*Mike Jerram*]

RIGHT The guns aren't real but *Sally B*, the only airworthy B-17 in Europe, looks convincing thanks to a television company which financed some aluminum-and-fiberglass turrets for a movie role. This B-17G saw no active war service, but flew as a high altitude photo-ship for the French government's mapping service until the Eighth Air Force Memorial Flight acquired it as a tribute to the 79,000 American airmen who died over Europe in World War 2. [*Mike Jerram*] [*Photograph below by Michael O'Leary*]

RIGHT The B-17's flight deck is nothing if not cosy! But all controls fall easily to hand and visibility through the windscreen is excellent. During steep turns (e.g. on tight finals or evasive action), the three-quarter roof panels above the flight deck were invaluable. The navigator and bombardier sat below and in front of the captain and co-pilot, well into the nose. [*Norman Pealing*]

BELOW Once upon an airshow there were two B-17s flying in Britain. Here, *Sally B* formates behind the sister ship now on permanent display in the Bomber Command hall at the Royal Air Force Musuem, Hendon. Some of its low-life parts and 1200 hp Wright Cyclone R-1820-65 nine-cylinder radials were exchanged to keep *Sally B* in flying trim. Based at Duxford, *Sally B* currently maintains a high-profile on the UK airshow circuit 43 years after she rolled off the Lockheed-Vega line at Burbank. The aircraft has worn many different schemes since arriving in the United Kingdom from France in 1975. Wearing the overall polished silver scheme synonymous with USAAF in 1944–45, *Sally B* toured the circuit for many aircraft for many seasons before emerging from its winter hide at Duxford several years ago in Olive Drab, its tail bedecked in 41st Combat Bomb Wing markings. [*Norman Pealing*]

ABOVE For old-timers pictures like this and the one below will bring back powerful memories of the mass daylight raids launched from bases in East Anglia and the South-east of England during World War 2. The B-17G was the most heavily armed bomber of the war, but all the guns in the world and powerful fighter escort could not prevent many Fortresses falling to the often deadly accurate flak put up by the German defenses. [*Norman Pealing*]

Heavyweight champion 2: LIBERATOR

Built in far greater quantities than even the B-17 Flying Fortress, the mighty B-24 Liberator was the most mass-produced bombing aircraft in history and the most numerous of all American warplanes: production ended on 31 May 1945 at number 19,203 – excluding another 1800 equivalent aircraft built as spares.

The tireless men and women on the Ford and Consolidated production tracks turned out not only 10,208 bomber versions (the B-24G, H and J), but LB-30A and C-87 transport versions, C-109 fuel tankers to supply B-29s in China, TB-24 trainers, CB-24 lead ships and F-7 photo-reconnaissance aircraft.

Yet the Liberator has always received less than its fair share of publicity or fame and today the B-24 is one of our rarest warbirds, a handful surviving in the world's museums.

LEFT The massive snout of the B-24J owned by Yesterday's Air Force – the only true bombing version of the Liberator still flying. David Tallichet, President of YAF, was well aware that the Indian Air Force had, amazingly, been operating Liberators on ocean patrol duties until the late 1960s. The Indians, instead of scrapping all the aircraft (which were replaced in their duties by almost equally aged Constellations), made some of the airframes available to museums. Pima County Museum in Arizona received an example, as well as the RAF Museum and Canadian War Museum, but none of these aircraft were destined to fly again. Tallichet had other plans and had his B-24J, registered N94459, prepared for an epic flight back to America. [*Michael O'Leary*]

B-24 LIBERATOR

BELOW In an historic flight Liberator N94459 (s/n 44-44272) was flown to Duxford, England, and given lavish care by the Duxford Aviation Society. From there the bomber headed towards America (a collapsed nose gear delayed the journey) and eventually arrived at Tallichet's main base in Chino, California. While at Duxford, the plane had been highly polished (the Indians had left their Liberators in remarkably good and original condition – all the turrets were still in place), but the fortunes of the veteran were to go downhill after arriving home. The Liberator attended several airshows and then slowly lapsed into a non-flying condition, though several attempts have been made to get the four-engine bomber back into the air. The most recent trip saw the bomber travel from March AFB to Liberal, Kansas, where it was forced down due to engine problems.

The plane, pictured here at Chino in September 1979, is currently being repaired and made airworthy to become the only original B-24 bomber flying. [*Michael O'Leary*]

INSET LEFT The B-24J's massive twin tails and rear turret are highlighted in this close-up. [*Michael O'Leary*]

INSET RIGHT This view shows the waist gun position with its enclosure attached. Note the wind deflector to help the gunner turn his .50 caliber weapon more easily in the slipstream. [*Michael O'Leary*]

RIGHT The Confederate Air Force's LB-30B transport, *Diamond 'Lil'*, is steadily being converted to B-24 bomber configuration. [*Norman Pealing*]

ABOVE The earliest 'Libs' were judged too immature for combat operations in Europe, so the type was initially employed by the RAF Atlantic Return Ferry Service as LB-30A transports in the spring of 1941. This trailblazing led to the Liberator I for RAF Coastal Command, which was equipped with top secret ASV radar and armed with fixed 20 mm Hispano cannons in the nose. Later, PB6Y-1s (RAF: Liberator IV) of the US Navy and Coastal Command made a tremendous contribution to the Allied victory in the Battle of the Atlantic by sealing the mid-Atlantic patrol gap, previously an important haven and assembly area for 'wolf packs' of German U-boats. The B-24C bomber entered service with the US Army in November 1941, shortly before the Japanese pre-emptive strike on Pearl Harbor. [*Norman Pealing*]

ABOVE RIGHT As its long span, high-aspect ratio Davis wing implies, the B-24 Liberator was designed as a long-range, high-speed bomber for the USAAF. The XB-24 prototype made its maiden flight on 29 December 1939. After the war many Liberators were converted for passenger use, especially in Latin America. [*Norman Pealing*]

RIGHT Delivered to the RAF as AM927, *Diamond 'Lil'* is the world's oldest surviving Liberator, being the 24th off the production line. After the war it was rebuilt by manufacturer Consolidated as a decidedly roomy corporate transport. An original Liberator nose section (almost certainly from a Navy PB4Y) has already been installed and the current 'civil' rear end is next on the list of modifications. Anyone have a spare Consolidated or Motor Products tail turret? [*Norman Pealing*]

Heavyweight champion 3: SUPER-FORTRESS

The first pre-production YB-29 made its maiden flight on 26 June 1943 and it was easily the world's most advanced aircraft, representing a quantum leap in a whole range of technologies including manufacturing techniques, structures, airborne systems, pressurization, engines, armament and wing loading.

The Superfortress proved beyond doubt that strategic bombing could win wars by neutralizing the enemy's industrial and military capabilities. Even before the specially modified Martin-built B-29s *Enola Gay* and *Bookscar* of Colonel Paul Tibbets' 509th Composite Group delivered the atomic bombs 'Fat Man' and 'Little Boy' over Hiroshima and Nagasaki to end World War 2, massed formations of up to 500 B-29s had laid waste to Tokyo and other Japanese cities and manufacturing centres with a devastating series of fire raids.

By VJ-Day over 3000 Superforts had been delivered by a production organization comprising Boeing, Bell, North American, Fisher (General Motors) and Martin.

RIGHT Perhaps the most impressive of all the aircraft preserved by the Confederate Air Force is *Fifi*, their magnificent Boeing B-29A Superfortress N4249 (s/n 44-62070, c/n 11547), which is kept in fine flying condition. [*Norman Pealing*]

FAR LEFT Access between the flight deck and fire control center amidships is by way of a pressurized tube – just big enough to crawl through – which passes over the bomb-bays. The fire control system was a radical advance and was entirely remotely controlled. Subsequently, hundreds of B-29s had their four General Electric twin .50 caliber machine-gun turrets removed to increase speed and altitude performance, leaving only the gunner in the Bell tail turret (equipped with one 20 mm cannon and twin .50 calibre guns or three .50 calibre guns) for self-defense. [*Norman Pealing*]

LEFT Powered by four 2200 hp Wright R-3350-23 Duplex Cyclone 18-cylinder radials with exhaust-driven turbochargers, the B-29 cruised at 290 mph and had a range of 3250 miles with a 10,000 bomb load. Initially B-29s raided Japan from bases in India and China, but they were able to operate from the Saipan, Guam and Tinian islands when these were captured from the Japanese in 1944. [*Norman Pealing*]

B-29 SUPERFORTRESS

The B-29 could carry a maximum internal bomb load of 20,000 lb, but for 'Project Ruby', a joint American-British operation in 1946, three B-29s had their bomb-bays specially adapted for what we would today call the semi-conformal carriage of a British 22,000 lb Grand Slam 'earthquake' bomb. Designed by the great Dr Barnes Wallis, this supersonic penetrator had been used with spectacular effect by the Lancasters of No 617 (Dambuster) Sqn near the end of the war, notably against the Bielefeld Viaduct in Germany. The modified Superfortresses successfully 'Grand Slammed' the previously invulnerable

German submarine assembly plant at Farge and other similar hardened structures. None of the 88 B-29s supplied to help the RAF in 1950–58 (in which service they were known as Washington B.1s) were ever modified to carry Grand Slam.

A particular characteristic of the B-29 is its flat climb out after rotation – not to be mistaken for a laggardly rate of climb – portrayed here as *Fifi* takes off in style from Harlingen for a re-enactment of the end of the Pacific war during the 1987 Confederate Air Force display. The huge Fowler flaps are electrically driven. [*Norman Pealing*]

BELOW The exceedingly roomy flight deck was a feature much appreciated by tired B-29 crews after ten hours or more on board. Access is via a ladder set up in the nosewheel bay. On the floor between the pilots is a transparent panel to check nosewheel centering. The lady is unforgiving if you neglect to straighten and lock the nosewheel leg before take-off. [*Norman Pealing*]

BELOW With his back to the co-pilot the flight engineer faced a station on the B-29 which set a standard followed by many a postwar airliner and bomber, jets included. The layout in the B-29 was the first time that the correlation between the duties of the pilots and flight engineer had been thought through on a calculated basis. The result was a reduced workload for all concerned, with better in-flight trouble-shooting and improved overall efficiency. [*Norman Pealing*]

ABOVE *Fifi* was restored from one of the many derelict B-29s which languished in the desert near the China Lake weapons range in California. Prepared for the flight to Harlingen in just nine weeks, the Superfortress made her first flight for 17 years on 2 August 1971 and, 6 hours and 28 trouble-free minutes after leaving China Lake, she touched down at CAF headquarters. [*Norman Pealing*]

LEFT Silhouetted beneath *Fifi*'s tail bumper and lower rear remotely-controlled gun turret is *Diamond 'Lil'*, believed to be the last airworthy Liberator. Behind her is a once-in-a-lifetime collection comprising a B-26, an A-26 and a B-25, plus a host of others lined up on the Harlingen ramp. One of the marvelous attributes of the 'guys and gals' who constitute the CAF is their total hospitality and friendliness. They come not only to admire the warbirds of other members but to take a trip in them, too. In August 1987 this photographer had a glorious 40 minutes in a CAF B-17 with the president of Southwest Airlines. [*Norman Pealing*]

RIGHT The Superfortress was destined for the Pacific theater where long-distance missions were the norm, and the great bomber was not deployed in Europe until after VJ-Day. [*Norman Pealing*]

BELOW Rescued from dozens of Superfortresses at China Lake Naval Air Station, where they were being used as targets for weapons testing, N4249 has been kept in operational trim since being obtained by the CAF. As well as attending many airshows, the veteran bomber has appeared in movies and television programmes, taking a lead role in the film *The Right Stuff*. [*Micheal O'Leary*]

BELOW Visibility from the B-29's cockpit is excellent, especially on the approach. Its hemispherical shape forms the front end of the pressurized crew compartment, the rear of which features a pressure dome – just like today's jetliners. With a service ceiling of 31,850 feet the Superfortress was a high flyer, making it less vulnerable to interception by fighters and optimizing fuel consumption. [*Norman Pealing*]

BELOW One of the Superfortresses that made it out 'alive' from China Lake NAS is the example on display at the Imperial War Museum, Duxford, England. After initial restoration work was carried out on site, the big bomber was flown to Tucson, Arizona for further restoration before making the lengthy journey to Britain. Registered as G-BHDK, TB-29A-45-BN 44-61748 had seen service with the 307th Bomb Group and flew 105 missions over Korea – and is painted in appropriate markings. Unfortunately after its ferry flight the bomber was never to fly again. [*Michael O'Leary*]

ABOVE This is where the Confederate Air Force's Superfortress came from . . . the vast collection of B-29s assembled in the Mojave Desert for target practice. This forest of aircraft shows off a wide variety of markings from World War 2 to Korea. Fortunately, the dry desert weather kept the marginally undamaged bombers in good shape (many of the 200 plus Superfortresses flown to China Lake were blasted to bits) and eventually allowed two other B-29s to fly out. After an initial go by several museums in the 1970s many of the Superfortresses were scrapped and only two rather ragged B-29s remain at China Lake for inclusion in a planned China Lake NAS Museum. [*Michael O'Leary*]